W9-BZW-196

THE WONDERFUL WORLD OF MATHEMATICS

A CRITICALLY ANNOTATED LIST OF CHILDREN'S BOOKS IN MATHEMATICS

Second Edition

By

Diane Thiessen
University of Northern Iowa

Margaret Matthias
Southern Illinois University at Carbondale

Jacquelin Smith
University of Northern Iowa

NATIONAL COUNCIL OF
TEACHERS OF MATHEMATICS

Copyright © 1992, 1998 by
THE NATIONAL COUNCIL OF TEACHERS OF MATHEMATICS, INC.
1906 Association Drive, Reston, VA 20191-1593
All rights reserved

Library of Congress Cataloging-in-Publication Data:

The wonderful world of mathematics : a critically annotated list of
children's books in mathematics / [edited] by Dianne Thiessen,
Margaret Matthias, Jacquelin Smith. — 2nd ed.
    p. cm.
    Includes index.
    ISBN 0-87353-439-5
    1. Mathematics—Juvenile literature—Bibliography. I. Thiessen,
Diane. II. Matthias, Margaret. III. Smith, Jacquelin.
Z6651.W66   1998
[QA40.5]
016.51—dc21   98-15748
                                                           CIP

Cover design and all illustrations by Bill Firestone.
Photos by Pat Fisher and Holly Cutting Baker.

Printed in the United States of America

Contents

Preface to the Second Edition

Invitations to explore mathematics abound in the wonderful world of children's books in mathematics. The colorful covers of children's books are designed to entice prospective readers. The contents are written to delight, entertain, and inform. These features are enticements for children to become actively involved in constructing mathematics.

From the simplest to more-complex topics, these books can engage readers in worthwhile tasks. Our analysis of children's books and the ways in which they can be used has revealed a rich resource for developing mathematical power. These books can be appreciated by adults as well as by children and can be used on a one-to-one basis between parent and child, as a classroom exploration for a group of children, or for individual enjoyment. Through sharing these books, adults and children not only will be entertained but will develop confidence in their ability to engage in mathematical tasks involving problem solving and reasoning. And the books present an opportunity for children to communicate about mathematics.

In our work with preservice and in-service elementary school teachers, parents, and others, we have also found that children's literature in mathematics has been a valuable tool for developing positive attitudes toward mathematics as well as for exploring mathematics. The attitude of the reluctant mathematician can be changed through the exploration of quality mathematics books for children. In a highly technological society, developing positive beliefs and dispositions is essential.

The focus on process in mathematics is concurrent with a similar focus in the area of reading. Literature that explores mathematical concepts appears to be a natural tool for attaining goals in both these areas, particularly at the emergent and early reading levels. An awareness of the potential role of quality mathematics books for children led to the compilation of this resource for educators.

In the past decade, children's literature in mathematics has been popularized. Evidence of this can be found in—

- the number of resources on how to use children's books in the classroom;
- the number of conference sessions having children's literature in the description;
- the number of articles on children's literature in *Teaching Children Mathematics* and the creation of the column "Links to Literature";
- the number of new children's books in mathematics.

Children's books in mathematics are being used more frequently in classrooms. Articles and books that illustrate how to use children's books have provided good models for incorporating literature into the mathematics classroom. A number of these models reflect worthwhile tasks, classroom discourse, and learning environments as described in the *Professional Standards for Teaching Mathematics*. Some teachers report that such resources gave them their start with using literature to support mathematics learning, and they now competently and confidently use new children's books as they become available. These teachers also report that they continue to look for new books to share with their students.

Both the quantity and the quality of children's books in mathematics has changed. There are more books that deal with a greater range of topics as well as more sophisticated topics. Edens's *How Many Bears?* is an excellent example of a book that combines counting, operations, and problem solving with stunning illustrations and a clever format. Pinczes's *One Hundred Hungry Ants* involves divisors of one hundred but can readily be extended to discussions about divisibility, primes, and composites. Scieszka's *Math Curse* delights readers of various ages; some of the mathematics will not be accessible to the young reader, yet the twists and humor of the book can be appreciated by all. These books and others will be read and reread and should motivate children to seek other books and to pursue other topics in mathematics.

Throughout the review process, we noted that children's books in mathematics are dominated in number and quality by the counting book. Faced with more than two hundred counting books, we were forced to examine the possible reasons for this dominance. The importance of beginning reading and the acceptance of counting as the first step toward number sense combine with the search for appropriate rhythmic and predictable text for emergent readers. The combination of familiar counting activities, a simple story line, and appealing illustrations provide children and their parents or teachers with ideal introductions to both mathematics and the reading process. The supply, understandably, responds to the demand. The dominance of the counting book may also reflect the restricted societal view that mathematics is limited to number and that the exploration of more-complex mathematical concepts is only for the few. Changing these attitudes is a constant challenge. We can only hope that as our readers join us in our journey of discovery, they will contact bookstores and publishers to place their orders for the books that have offered them the most insight into the excitement and challenge that the world of mathematics offers.

The Selection of Books

More than five hundred fifty books are reviewed in this second edition. More than 60 percent of these titles are new. These books are trade books, not workbooks or teacher-resource materials. The 1996–97 and 1997–98 editions of *Children's Books in Print* were the main resources for compiling and updating a book list. Another print source was *Forthcoming Books*. Additionally, recommendations from teachers and other children's-literature enthusiasts were considered.

The books reviewed in the first edition were included in the second edition if they were in print according to *Children's Books in Print*. This choice is arbitrary because the publishing status of books changes daily.

A quandary in book selection was deciding which books had mathematics concepts as a primary emphasis and which books contained mathematics that was incidental. Many books were clearly mathematics-concept books. The line dividing these books from books in which mathematics was incidental was not clearly defined. A number of the latter books can be a springboard to exploring different concepts or a catalyst for integrating different topics. This bibliography contains the books that are primarily mathematics.

A new section titled "Series and Other Resources" was created to accommodate the large number of books that have been issued in series. Because the books in each series have common elements, it is helpful to have all the annotations in one section. Each book in a series was cross-referenced in the appropriate topics for readers who are searching for books that include an explicit topic.

"Incidental Geometry—Quilting" is a new subsection. It was created because of the popularity of using children's literature whose main focus is quilting. Many teachers find these books invaluable in helping students make connections among geometry—in particular, tessellations—history, and quilting.

Some of the quality out-of-print books are listed in the introduction to various subsections. A number of these books can be found in your school or public library; a few may still be found in bookstores. All these titles were recommended or highly recommended in the original edition of *The*

Wonderful World of Mathematics. Some of these books were written by familiar authors. Since some of these books explore topics that are not encountered in other in-print books, we have included them here. A number of these selections were originally published as part of the Thomas Y. Crowell Young Math Books series.

The Crowell Young Math Books are notable for several reasons. Important mathematics topics are represented. A mathematics editor coordinated the series. Both activities and illustrations were carefully designed to involve the reader in exploring and developing concepts. The books are appealing. Often the publisher's recommended grade levels appear to be a mismatch with the curriculum; the books were designed to introduce children to more-sophisticated mathematics rather than simply to arithmetic. They were also designed so that young children have an opportunity to *play* with important mathematical ideas that are studied more formally in later grades. The presentation and quality of some tasks, such as those found in Froman's *Angles Are Easy as Pie* or Srivastava's *Spaces, Shapes, and Sizes,* have not been found in current books. For these reasons, we consider these books to be "classics" in juvenile mathematics literature, and although they are out of print, they should be acknowledged. Some of these books are also highlighted in vignettes from the classroom.

The Format of the Reviews

Each review describes the book's content and accuracy, its illustrations and their appropriateness, and the author's writing style and indicates whether activities for the reader are included. Numerals rather than word names are often used in the annotations to reflect the treatment of numbers in the books. We strived to have annotations that are consistent but at the same time reflect the uniqueness of each book.

In the bibliographical entries, a notation indicates whether the book develops a single concept or multiple concepts. The books are classified as listed in the table of contents. Some books are cross-referenced under more than one category. An overview for each of the five main categories can be found at the beginning of each section. More-detailed descriptions of the types of books are included at the beginning of each subsection. Vignettes in the overview and some subsections give a flavor of how some of the books have been successfully used in classrooms. Situations from elementary school, junior high school, and college classrooms illustrate how students of different ages can enjoy and use the books. By exploring appropriate books in the college classroom, preservice teachers can extend their knowledge about resources, teaching activities, and children's thinking all at the same time.

Each book is rated according to its usefulness in teaching mathematics concepts as highly recommended (★★★), recommended (★★), or acceptable (☆). Books that are not recommended for mathematics concepts have no designation. The grade level (from preschool [PS] to grade 6) of each book as stated in *Children's Books in Print* is included in the bibliographical entry. On the basis of our experiences with children, we questioned some of the recommended grade levels. For these books, our recommendations on grade levels are recorded in parentheses after the publisher's recommendation.

The ISBN number and the cost of the books, if available, are also included in the entries. Because prices fluctuate, the prices cited should be considered estimates. Some of the books are available with paper covers or with reinforced, or library, binding; this information has been recorded next to the appropriate ISBN number.

We hope the format and the information in the reviews will provide a sufficient basis for the intelligent selection of exciting books that will encourage explorations in mathematics for children in school and at home. We wish you happy reading and growth in mathematics prowess.

DIANE THIESSEN
MARGARET MATTHIAS
JACQUELIN SMITH

Acknowledgments

For the first edition of *The Wonderful World of Mathematics,* a committee was formed to read and review books. Parts of the second edition reflect their work. This committee included the following:

Nancy S. Angle, Cerritos College, California

Marilyn Bieck, Encinitas School District, California

Audrey V. Buffington, Wayland Public Schools, Massachusetts

Mary Jo Cliatt, University of Mississippi

Lynn Columba, Lehigh University, Pennsylvania

Kay Dean, Northern Arizona University

F. M. (Skip) Fennell, Western Maryland College

Cathy Grace, Southern Association on Children under Six

Ann Harsh, Hattiesburg Public School District, Mississippi

Audrey Jackson, Claymont Elementary School, Chesterfield, Missouri

Joanie Janis, University of California

Eula Ewing Monroe, Western Kentucky University

William Nibbelink, University of Iowa

Barbara Reed, Waverly Community Schools, Lansing, Michigan

Merrilyn Ridgeway, Arizona Department of Education

Nina L. Ronshausen, Texas Tech University

Sally Schneider, St. Raphael Elementary School, Louisville, Kentucky

Susan Sherwood, Wartburg College, Iowa

Rose Strahan, Delta State University, Mississippi

Timothy J. Touzel, University of South Carolina

Sharon L. Young, Palm Harbor, Florida

The authors thank especially—

- Lucille Lettow, University of Northern Iowa Youth Librarian, for her invaluable counsel in recommending books, locating book lists, and offering advice and encouragement throughout the project;
- the publishers who graciously responded to our requests for review copies.

Early Number Concepts

W ITH more than 200 counting books from which to choose, we have again decided to organize this large category of books in as user-friendly a way as possible. For the first edition of *The Wonderful World of Mathematics* we read and reread the entire counting category, looking for patterns and trends that would allow us to subcategorize. We became aware of the many opportunities that any one of these books could offer in the hands of a creative teacher or other adult. Each one could be used to emphasize processes in the development of problem-solving skills and mathematical reasoning. As we used these books with youngsters, our experiences have taught us how difficult it would be to avoid communicating the mathematical concepts included. Developing meaning for numerals and number words is central in most of the simplest counting books, and frequent exposure to these books cannot help but familiarize children with the beginning language of mathematics. We were also excited to find how many of these misleadingly simple books made wonderful stepping-stones or connections from number to other, more complex mathematical concepts.

The primary intent of practically all the books in this category is to give opportunities for counting. Each counting book is unique, but as you read more, certain characteristics across books will be noted. Most books are organized around a common theme. Some books have a definite story line, which is told by pictures or text. Practically all books involve rational counting, but since the sets are usually presented in sequence, readers could rely on rote counting as they proceed from page to page. By verifying the appropriate number of objects on a page, children are involved in rational counting experiences. When learning a concept, children need to build connections among the models, symbols (numerals), and word names, as shown in figure 1. Most books include models, but not all include both numerals and word names. Some books include counters in one-to-one correspondence to the objects depicted on the opposite page. The child's experiences with these books can be extended by relating counting to other situations in the young reader's environment.

We have classified the counting books into ten categories. Each category is preceded by a short explanation to help clarify its potential use. It

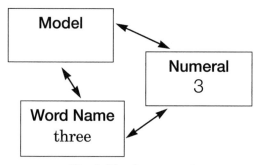

Fig. 1. Number concepts

should be noted that the classification system we have devised for the counting books, or any other major category, is not perfect. Placement in one category does not mean that a book may not also belong to another category. We have tried merely to identify the topic with which each book has the closest relationship. The categories of counting books are "Counting—One to Ten"; "Chants, Songs, and Rhymes"; "Number in Other Cultures"; "Counting—Zero"; "Counting—Large Numbers"; "Counting— Place Value"; "Counting—Bar Graphs"; "Counting—Partitions"; "Counting—Operations"; and "Counting—Other Number Concepts."

As described below, many conversations and concepts can emerge from sharing a single book.

Four-year-old Sara *loved* Ryan and Pallotta's *The Crayon Counting Book*. In fact, she asked to have her own copy. Earlier, when Elana Joram was selecting books for her daughter, she expressed reservations about choosing counting books, since Sara was losing interest in them. Later Elana gave this report:

> The idea of counting in a different way was quite compelling to Sara and also to me—I had simply not thought of attempting to teach her anything other than the "regular" counting sequence. As we went through the even numbers, I first tried explaining to her the principle that was used to generate the numbers, and then tried to get her to make predictions about which number came next. Her predictions were quite off (for example, her prediction for what comes after 8 was 7— indicating at least that she knew this was not regular counting) but she seemed to enjoy trying to generate them. What she liked the most were the pictures—at one point she said, "This is magic!" She explained, "because first we didn't have very many, and now we have a lot and we didn't even do anything." The next morning I found her reading the book by herself on the couch, and as I said, she has been quite obsessed with it.

As a parent, I find that this book seems to have a lot of potential to be used repeatedly over a long time. It has opened my eyes to the idea that I can present Sara with some interesting and unusual things she can do with numbers. I have had a real prejudice against books that focus on conceptual knowledge rather than books that present concepts in a story line, but seeing Sara's reaction to this book has made me become aware of and rethink this bias.

Sara's interest in odds and evens has continued. She says she likes odds better, but I think this has to do with the colors of the crayons in *The Crayon Counting Book*. One day we played a game of guessing the next number as we reread the book. The idea of counting by twos connected nicely with Hamm's *How Many Feet in the Bed?* She also became quite interested in zero and looked for examples of it frequently in her daily life. For example, according Sara, her eleven-week-old cousin, Benji, is "zero," not even "one." When I responded that yes, he is zero years but he is eleven weeks old, she looked puzzled.

Counting—
One to Ten

Since counting is the initial experience for the youngest readers in developing number concepts, it seemed appropriate to place the simplest books, which primarily focus on counting one through ten or fewer, in a category of their own. Within this first group of books are many opportunities for the young reader to develop a growing understanding of number. Authors have chosen a number of ways to motivate children to count. The story form of Och's *When I'm Alone,* the rollicking rhyme of Calmenson's *One Little Monkey,* and the mystery provided by the manipulative design of Van Fleet's *One Yellow Lion* are only a few examples of the variety that awaits eager young readers in this category. Also motivating is the variety of quality illustrations: large color photographs dominate *The Lifesize Animal Counting Book;* charming, whimsical illustrations are found in Tryon's *One Gaping Wide-Mouthed Hopping Frog;* classical art work is introduced in Voss's *Museum Numbers,* illustrated by Kyung Mi Ahn. Some books, such as Bang's *Ten, Nine, Eight,* introduce counting backward. Numerals and number words are reinforced in many of the books; Blumenthal's *Count-a-saurus* and Ruschak's *The Counting Zoo: A Pop-Up Number Book* are two examples. Finally, Hoban's *1, 2, 3* places the counting experience in the child's everyday life.

Other books that feature counting from one through ten are also included in "Chants, Songs, and Rhymes"; "Number in Other Cultures"; "Counting—Bar Graphs"; "Counting—Operations'; and "Counting—Other Number Concepts."

The following books are out of print but are considered exceptional:

Inkpen, Mike. *One Bear at Bedtime.*
Wood, A. J. *Animal Counting.*

Copies may be available in your library. See the first edition of *The Wonderful World of Mathematics* for annotations.

Arkadia. *Learning with Peter Rabbit: A Book of Numbers.* Avenue, N.J.: Derrydale Books, 1993.

ISBN 0-517-07697-7 $ NA

☆ Single concept, PS

Based on the stories and art of Beatrix Potter, this small board book introduces the numbers from one through ten and their corresponding numerals. Each numeral is printed in large type, and the word name, incorporated in the story line, is emphasized in bold uppercase letters. Peter Rabbit and his sister, Cottontail, are the two main characters in the simple text that revolves around their romp in Farmer McGregor's garden. As they explore, they jump over three flower pots, wake up four mice, and gobble eight onions. An appropriate number of objects to be counted are illustrated in familiar Potter style and convey the charm of the much-loved characters.

Ashton, Elizabeth Allen. *An Old-Fashioned 1-2-3 Book.* Illustrated by Jessie Willcox Smith. New York: Viking, 1991.

ISBN 0-670-83499-8 $14.95

★★ Single concept, PS–3 (all ages)

Paintings by Jessie Willcox Smith, who illustrated such children's books as *Heidi* and *Little Women* over fifty years ago, are the focus of this counting book. The author carefully selected Smith's work to illustrate the numbers from one through ten. An illustration of two girls looking at their reflection is introduced by Ashton's verse on the opposite page: "Heather and Jill, Jill and Heather / Kneel upon the grass. / Two new friends make four together / In a lake's looking glass." A stylized numeral 4 appears on the same page; all pages are framed in a border of blue scrolls to complete an old-fashioned look. The illustrations are charming and well presented. Some sets are difficult to count; in particular, one of the nine children is partially hidden. In the illustration of ten dolls, it is difficult to determine distinct dolls because most are partial pictures. The old-fashioned illustrations are more likely to appeal to adults, but this well-crafted book is a good introduction for the young reader to this classic art.

Astley, Judy. *When One Cat Woke Up: A Cat Counting Book.* New York: Dial Books for Young Readers, 1990.

ISBN 0-8037-0782-7 $10.95

☆ Single concept, PS–2

Colorful illustrations of a home and a simple text trace the mischievous antics of one cat who wakes from a nap. She steals two fish out of the refrigerator, fights with three teddy bears that had decorated a bed, crumples four shirts that were hanging in a closet, ..., and finally leaves ten muddy paw prints on her way back to sleep in a laundry basket. The pleasing, primitive illustrations capture the charm of the home. For example, the three teddy bears are shown lying on the quilt covering a brass bed; toys, including trains and stuffed animals, indicate a child's bedroom. A border surrounds the opposite page, which pictures the cat and the three torn teddies. The text includes the number word name, and a large numeral 3 dominates the upper right-hand corner. The end pages vertically list the numerals from 1 through 10 across from illustrations of one cat, two fish, three teddy bears, and so on.

Bang, Molly. *Ten, Nine, Eight*. New York: Greenwillow Books, 1983.
ISBN 0-688-00906-9 $16.00
ISBN 0-688-00907-7   (library binding)   14.93
ISBN 0-688-10480-0   (paper, Mulberry)   3.95
ISBN 0-590-7313-3   (Scholastic)   19.95
★★ Single concept, PS–1

Ten, Nine, Eight is a colorfully illustrated countdown to bedtime, featuring a little African American girl and her father. A very loving atmosphere is created through inviting illustrations and rhythmic, poetic words. Some of the counting opportunities are events rather than pictures, such as "3 loving kisses on cheeks and nose." Counting from ten through one is the primary objective.

Benjamin, Alan. *Let's Count Dracula*. Illustrated by Sal Murdocca. New York: Little Simon Book, 1992.
ISBN 0-671-77008-X $3.95
☆ Single concept, PS–K

Number concepts and numerals from 1 through 8 are introduced in this small, cardboard-paged, square-shaped "Chubby Board Book." A Halloween theme is reflected in each detailed, double-paged illustration and in the simple text, which begins with "1 moon, shining / 2 black cats, whining." Most of the sets are easy to count; in one instance, young readers will need to search carefully for the "5 witches, hiding." The theme is completed as Dracula appears on the page depicting "7 spooks, meeting" and a line of costumed children appear on the final page of "8 kids, trick-or-treating."

———. *Zoo Numbers*. New York: Little Simon, 1994.
ISBN 0-671-86601-X $3.95
☆ Single concept, PS–K

This sturdy board book features only numerals as the young reader goes on a photographic safari through the vicarious visual experience provided by the magnificent photographs from the Zoological Society of San Diego. The close-ups allow easy counting of various groups from 1 through 10, including 1 elephant, 2 rhinoceroses, 3 zebras, and 7 Przewalski's horses. Unique photographs include a group of 8 hatchlings at various stages of feathering and a composition of 9 starlings lined up on a curved branch with a great egret below. The book will catch a toddler's eye and encourage him or her to label and count. The end pages recap what has been counted and name each animal group.

Bishop, Roma. *Numbers.* New York: Little Simon, 1991.

ISBN 0-671-74832-7 $2.95

☆ Single concept, PS–K

Numerals from 1 through 6 pop up as each sturdy cardboard page is turned in this tiny book. The only text is the appropriate number word name, which appears on each page. The objects—one clown, two socks, three balls, four apples, five ducks, and six butterflies—are distinctly pictured so they are easy to count. The illustrations as well as their backgrounds are done in cheerful bright colors that should appeal to the preschooler.

Blumenthal, Nancy. *Count-a-saurus.* Illustrated by Robert Jay Kaufman. New York: Simon & Schuster Books, 1989.

ISBN 0-02-749391-1 $13.95
ISBN 0-689-71633-8   (paper)   3.95

★★ Single concept, PS–3

Here is a tongue twister all youngsters can enjoy. *Count-a-saurus* explores ten species of dinosaurs in rollicking rhyme from "One Stegasaurus standing in the sun" to "Ten Hadrosaurids sporting funny hats." The double-page illustrations are hilarious, which adds to the light tone of this delightful book. Numerals and number words are given. A pronunciation guide on each page will help children with the difficult dinosaur names. Additional information about the featured dinosaur species can be found at the back of the book in an "append-a-saurus."

Bright, Robert. *My Red Umbrella.* New York: William Morrow, 1985.

ISBN 0-688-05249-5 $12.00

☆ Single concept, PS–1

Delicate black-and-white line drawings are interrupted only by the bright red splash of color provided by the umbrella, which grows to accommodate more and more animals. Sets of animals from one through four gather under the little girl's umbrella to shelter from the rain. There is no attempt in the simple text to highlight number words; the counting objective is only implied.

Calmenson, Stephanie. *One Little Monkey.* Illustrated by Ellen Appleby. New York: Parents Magazine Press, 1982.

ISBN 0-8193-1091-3 $5.95
ISBN 0-8193-1092-1   (library binding)   5.95

★★ Single concept, PS–3

Cumulative story form and rhyming text unfold to reveal the rollicking events of a jungle stampede caused when the animals misunderstand the reason for the monkey's dash through the treetops. The rhyme bounces along as two hippos, three zebras, four antelope, and so on, flee and the pages become more and more crowded with colorful creatures. Number concepts from one through ten are clearly presented by the expressive illustrations of the animals.

Carle, Eric. *1, 2, 3 to the Zoo.* New York: Sandcastle Books, 1990.
ISBN 0-399-21970-6 (paper) $6.95
★★ Single concept, PS–2

Colorful collages of a train full of lively zoo animals effectively introduce sets and numerals from 1 through 10. Each successive train car carries a different group, from one elephant, two hippopotamuses, three giraffes, to ten birds. Across the bottom of each page is a tiny silhouetted diagram of the entire train and the cargo that has been introduced. The last page shows the silhouettes of the empty cars, since the animals are in their new zoo environment. This is a reprint of a book originally published in 1968.

———. *The Very Hungry Caterpillar.* New York: Philomel Books, 1987.
See Measurement: Measurement of Time—Concepts

Carter, David A. *Baby Bugs Pop-Up Book: Counting.* New York: Little Simon, 1993.
ISBN 0-671-86876-4 $4.95
☆ Single concept, PS–K

An exciting book for the toddler, this pop-up book with sturdy cover and pages will captivate any young child. At each page turn, progressively more beautiful butterflies fly up before the reader's eyes, from one butterfly to six and then ... "Bugs galore." Number words are in uppercase letters and the appropriate numerals are shown. The illustrations and design of this tiny book make for easy counting.

———. *How Many Bugs in a Box?* New York: Simon & Schuster, 1988.
ISBN 0-671-64965-5 $12.95
★★★ Single concept, PS–2

Cleverly designed pop-ups introduce the numbers from one through ten in this sturdy book. On each double page the reader is asked, "How many bugs are in the (blue, green, tall, square, thin, small) box?" The adjectives well describe the variety of intriguing boxes the reader is to explore. Each pop-up is unique; the flaps fold down or up or to the left or pull out or up or to the right. Each number concept is illustrated with a unique set of colorful "bugs" and named with the corresponding numeral and word name, for example, "5 / Five / mellow / yellow / fish-bugs." There are surprises on each page, such as the "4 / Four / fast / fleas" that scurry out of sight as the flap is opened. The book is a delight for readers of any age.

Chandra, Deborah. *Miss Mabel's Table.* Illustrated by Max Grover. San Diego, Calif.: Harcourt Brace Jovanovich, 1994.

ISBN 0-15-276712-6 $14.95

★★ Single concept, PS–2 (2–3)

One frying pan, two teaspoons of salt, three glasses of milk to ten dashes of yeast are sequentially set on Miss Mabel's table. Each new ingredient is introduced in cumulative rhyme with the corresponding number of items appearing on the table: "that lie near the frying pan, big and bold, / waiting to cook on the kitchen stove, / but now, so quiet and empty and cold, / sits on Miss Mabel's table." On alternating pages, illustrations depict a red-haired woman starting her day in the darkness of 5:00 A.M. and commuting across the city on bicycle and public transportation. The text does not refer to this individual or to the purpose of the ingredients until the end of the story, when the young woman arrives at a restaurant, Miss Mabel's Table, and fixes pancakes for ten of the nearby residents and workers. Counting is a secondary theme; the reason for the ingredients and how the red-haired woman relates to the accumulation of the ingredients are the book's main focus. The illustrations, done in a primitive style with bold acrylics, give hints to the mystery. Various ethnic groups are represented as well as women in nontraditional roles.

Duerrstein, Richard. *One Mickey Mouse: A Disney Book of Numbers.* New York: Disney Press, 1992.

ISBN 1-56282-266-7 $5.95

☆ Single concept, (PS–K)

Mickey and his friends introduce the numbers from one through six in this sturdy, square-shaped board book. A simple text and a numeral announce each famous character, such as one Minnie. The opposite page features a smiling Minnie framed in a bright yellow square that is lightly speckled with black paint. This design is used throughout the book. Each speckled square is divided into the appropriate number of equal-sized panels with a cartoon figure posed in each section. Preschoolers should find this book appealing.

———. *Un ratón Mickey: Un libro Disney de números.* Translated by Daniel Santacruz. New York: Disney Press, 1993.

ISBN 1-56282-460-0 $5.95

★★ Single concept, (PS–K)

A welcome edition is this Spanish version of *One Mickey Mouse: A Disney Book of Numbers.* See the review above.

Edens, Cooper. *The Wonderful Counting Clock.* Illustrated by Kathleen Kimball. New York: Simon & Schuster Books for Young Readers, 1995.

See Measurement: Measurement of Time—Telling Time

Ernst, Lisa Campbell. *Up to Ten and Down Again.* New York: Lothrop, Lee & Shepard Books, 1986.

ISBN 0-688-04541-3 $13.95
ISBN 0-688-04542-1   (library binding)   13.88
★★ Single concept, P–K

Numbers from one through ten and down again are depicted in these pleasing illustrations pictured as photographs in an album. Each layout shows a "photograph" of the picnic scene with "snapshots" showing the appropriate numeral and each object to be counted. To represent five, the lush country scene shows five girls and separately frames each girl and the numeral 5 in six smaller snapshots. This format clearly shows the one-to-one correspondence between the two sets. When ten storm clouds appear, the children and adults start packing up and recounting the nine hats, eight baskets, seven boats.... Each scene is viewed by the duck who was serenely swimming on the opening page. The last scene shows the duck playing in the water, joined by other creatures. There is no text, but the detailed illustrations tell a charming story.

Estes, Kathleen. *The Mermaid's Purse.* Bridgeport, Conn.: Greene Bark Press, 1995.

ISBN 1-880851-24-5 $15.95
☆ Single concept, PS–1

Sealinda the mermaid and other fantasy sea creatures introduce the numbers from one through ten. As Sealinda continues to bargain with Erastus the clam for a large, perfect pearl, she offers to trade 1 jellyfish, 2 skates, 3 cowfish, ..., 9 sea horses. Each time, she removes one of the groups of fish from her purse; each time, the offer is rejected. When she suggests 10 sand dollars, the offer is accepted. Numerals are included with each appropriate set; a final page lists and shows each group. The young reader will be delighted with the skate fish that are skates; the cowfish that look like cows, and so on. The colors in the illustrations range from pastel to fluorescent. A glossary of sea creatures includes images of both the imaginary and real fish as well as written descriptions of the latter.

Garne, S. T. *One White Sail: A Caribbean Counting Book.* Illustrated by Lisa Etre. New York: Green Tiger Press, 1992.

ISBN 0-671-75579-X $14.00
★★ Single concept, PS–1

"One white sail on a clear blue sea, / Two orange houses and a slender palm tree." The rhyming story line introduces the numbers from one through ten. Each verse is set in large clear print in the center of each left-hand page, which is outlined with a simple, black line border. Each opposite page depicts a picturesque Caribbean scene painted in deep blues and greens, vibrant reds and oranges, and shades of lavender and pink. The stunning watercolor illustrations are most appealing and capture the colorful setting and culture. The objects to be counted are clearly depicted; only word names are introduced.

Green, Kate. *A Number of Animals.* Illustrations by Christopher Wormell. Mankato, Minn.: Creative Education, 1993.

ISBN 0-88682-625-X $19.95

★★ Single concept, PS–3

Engravings by Wormell dominate this simple counting book. A large black numeral introduces each set to be counted; word names are included in the text. The appropriate number of animals is shown on the opposite page as well as in miniature drawings underneath the numeral. The stark prints of farm animals shown in their environment reflect the descriptions expressed in the text. "Four fat turkeys ruffling their feathers / ... strut though the straw." The story line to introduce the numbers from one through ten has a lone lost chick inquire of two horses, three cows, four turkeys, and so on, if they have seen his mother. Ten chicks are shown as the lost chick is reunited with his nine brothers and sisters. A final page pictures the numerals from 1 through 10, with the appropriate set of farm animals underneath each numeral.

Grossman, Bill. *My Little Sister Ate One Hare.* Illustrated by Kevin Hawkes. New York: Crown Publishers, 1996.

ISBN 0-517-59600-8 $17.00
ISBN 0-517-59601-8   (library binding) $18.99

★★ Single concept, K–3

Hawkes's stunning fantasy illustrations set the stage for "my little sister's" performances that introduce objects from 1 through 10. Framed by the red theater curtain and standing in a bright spotlight, "little sister" first appears dressed as a magician; as she swallows a rabbit, the astonished audience of children appears to gasp. "My little sister ate 1 hare. We thought she'd throw up then and there. But she didn't." When appearing dressed as an Oriental snake charmer, she swallows 2 snakes; at a picnic she appears dressed in summer attire and eats 3 ants. The predictable rhyming text enumerates each previously swallowed item and ends with the same two sentences until the final scene. "My little sister," dressed in appropriate costumes, continues to shock her audience until she encounters 10 peas: "But eating healthy foods like this makes my sister sick, I guess." All of the outrageous scenes, illustrated in bold colors, are clever in design and bound to appeal to the humor of young children.

Halsey, Megan. *3 Pandas Planting.* New York: Bradbury Press, 1994.

ISBN 0-02-742035-3 $14.95

☆ Single concept, PS–2

Cheerful scenes in bright colors depict charming, nattily dressed animals in groups of twelve to one. Large black print announces the numeral and each group. For example, "12 Crocodiles" is the heading depicting carpooling; four jauntily appareled crocodiles are heading down the highway in a 1960s Chevy convertible; they pass three business-"diles" riding in an orange sedan and a family of five in a station wagon. The action of the animals will encourage an awareness of the many everyday activities that help clean up our world; saving water, recycling newspapers, organizing bottles for recycling, and planting trees are a few of the activities depicted. The message is strong and clear; "You can help us take care of the Earth." This simple countdown text includes added information for the more sophisticated reader, since each positive example is explained in more detail at the back of the book. Data abound on these final pages; for example, recycling all Sunday newspapers for one week would save half a million trees; fluorescent bulbs use 75 percent less electricity than incandescent bulbs do.

Hoban, Tana. *1, 2, 3.* New York: Greenwillow Books, 1985.

ISBN 0-688-02579-X $4.95

★★ Single concept, PS

Brightly colored photographs illustrate this small, sturdy, cardboard book. The numbers from one through ten are clearly represented by familiar objects: two red tennis shoes, three toy blocks, and seven animal crackers. This appealing counting book needs no text; the numeral, the word name, and the appropriate number of counters are printed in red on each page. Hoban's photographic genius continues to capture children's interest and stimulate their curiosity.

———. *Who Are They?* New York: Greenwillow, 1994.

See Geometry and Spatial Sense: Shapes

Hubbard, Woodleigh. *2 Is for Dancing: A 1 2 3 of Actions.* San Francisco, Calif.: Chronicle Books, 1995.

ISBN 0-87701-895-2 $13.95
ISBN 0-8118-1078-X (paper) 8.95

★★ Single concept, PS–3

2 Is for Dancing outpaces other counting books with its glorious celebration of color and action. Zany characters in appropriate motion accompany each action statement. From the relative calm of "1 is for Dreaming," the heat is turned on with "2 is for Dancing," "3 is for Jumping" to "10 is for Traveling." Clearly countable and clever are the six cats hanging from a pier with a seventh cat scuba diving for "7 is for Fishing" or the ten snails each with a black-and-white suitcase for "10 is for Traveling." Numerals and minimal text convey the concept well. Although counting is listed as the primary theme, the artwork and the exploration of actions dominate. A bonus is Hubbard's *C Is for Curious: An A B C of Feelings,* which is also included. When this book is turned over, the same vibrant colors and lively characters celebrate a superbly successful and alphabetic explorations of feelings, from angry to zealous.

Hutchins, Pat. *One Hunter.* New York: Greenwillow Books, 1982.

ISBN 0-688-00614-0		$16.00
ISBN 0-688-00615-9	(library binding)	14.93
ISBN 0-688-06522-8	(paper)	4.95

★★ Single concept, PS–1

One hunter bravely stalks his game in this simple counting book. He stealthily walks past two elephants, three giraffes, four ostriches, ..., and ten parrots, totally unaware of their presence. Imagine the children's delight when the animals begin to stalk the unsuspecting hunter. On one double-page layout, he quietly walks past a set of partially hidden giraffes while two elephants pop up behind his path. As he continues past a set of partially hidden ostriches, the three giraffes emerge from hiding. At the end of the book, the hunter shows his surprise as he turns around to encounter all the animals he has been seeking. Children will enjoy predicting each anticipated set.

Liebler, John. *Frog Counts to Ten.* Brookfield, Conn.: Millbrook Press, 1994.

ISBN 1-56294-436-3		$14.40
ISBN 1-56294-698-6	(paper)	3.95

Single concept, K–3

Frog is not allowed to play hide-and-seek with the other animals because he does not know how to count to ten. As this one frog takes a ride on his two-wheeled bicycle at three miles an hour, he becomes distracted by four flies and falls off a five-foot drop. The story line continues as the reader follows the frog's adventures as he inadvertently learns how to count to ten. A numeral is carefully incorporated into each pastel illustration; that is, the numeral 2 is the shape of the frog's arm; the numeral 6 is the path through which the falling frog is projected. Some sets are countable, like the four flies, but in other situations, like the five-foot drop, there is nothing to count. The focus is counting in order and finding the hidden numeral shapes. On the final page, the numerals from 1 through 10 are given in order as the frog shows his friends that he can rotely count. *Frog Counts to Ten* is a contrived story to encourage rote counting to ten; in contrast are the many quality books available that help the young reader develop counting skills *and* number concepts.

The Lifesize Animal Counting Book. New York: Dorling Kindersley, 1994.
ISBN 1-56458-517-4 $12.95
★★★ Single concept, PS–1

Captivating photographs of animals are effectively presented against a white background on oversize pages. Each close-up appears almost life size, since the photographs of the individual animals often extend over the red, blue, or green line border of each double page spread. Numbers from 1 through 10, 20, and 100 are illustrated by various groups of animals. Large black numerals and the number word names are included as well as simple descriptions, such as "Two contented cats" and "Seven snuffly rabbits." Some illustrations reflect addition concepts, such as "Three white ducks and five fluffy ducklings make eight." Different insects are alternately displayed and arranged in clearly defined rows for ease in counting the "One hundred creepy crawlers." Across the top border of each page, the appropriate number of ladybugs is shown; for illustrating the final double page, 100 ladybugs are evenly spaced around the border. The end papers show an array of 20 by 32 ladybugs that may evoke estimation or questions of how many. Simply presented, beautifully photographed, successfully focused, and totally appealing, this is an excellent choice for the preschool or primary school child.

Matthias, Catherine. *Too Many Balloons.* Illustrated by Gene Sharp. Chicago: Children's Press, 1982.
ISBN 0-516-03633-5 (library binding) $10.95
ISBN 0-516-43633-3 (paper) 3.50
★★ Single concept, PS–2

Predictable text is the vehicle through which a little girl tells of her visit to the zoo. On each page she buys balloons: first one red, then two yellow, then three blue, ..., and finally ten polka-dot. She shows each purchase to a different set of animals whose number is equivalent to the number of balloons. "I bought three blue balloons. I showed my three blue balloons to the seals. They liked them." The surprise ending will delight the young reader. A word list is included at the end of this well-written book, which is one in a series of "Rookie Readers."

Mayer, Marianna, and **Gerald McDermott.** *The Brambleberrys Animal Book of Counting.* Honesdale, Pa.: Bell Book, 1987.
ISBN 1-878093-75-4 (paper) $3.95
★★ Single concept, PS+

Introductory and end pages feature the numerals from 1 through 10, a tiny balloon high in the background, and two tiny, wistful mice looking in its direction. This foreshadowing is quickly understood as a runaway balloon leads Mrs. Brambleberry, a mouse, and her baby on a wild chase through the zoo. Movement through the story is emphasized by the presence of the balloon or its streamers floating off the right-hand page, which tempts the young page-turner to follow. Minimal text consists of numerals and the naming of the objects to be counted: one balloon, two mice, three birds, four boats, ..., ten bananas. Separations on the page offer opportunities for addition groupings; for example, there are five monkeys on one side of the page and four on the other. An interesting ending to the zoo visit occurs when ten bananas are available for nine monkeys. The problem is easily solved by the zookeeper bear, who eats the tenth one. A scene of utmost satisfaction brings the little adventure to a close with Mrs. Brambleberry and her little one, with balloon in hand, silhouetted against a glowing sunset. The pastel illustrations are soft and appealing. McDermott, a Caldecott Award winner, shows his talent in adapting his awesome potential to suit this simple text.

Morozumi, Atsuko. *One Gorilla*. New York: Farrar, Straus & Giroux, 1990.

ISBN 0-374-35644-0 $15.00

★★ Single concept, PS–2

Morozumi's lush illustrations were awarded the *New York Times'* Best Illustrated Children's Book Award. Each rich scene depicts a gorilla with a different number of animals from one through ten. The reader will need to search carefully to find all three budgerigars, five pandas, eight fish, and others, which are skillfully painted into detailed scenes; at times, they are partially camouflaged in their environment. Simple text introduces the one gorilla and the others in this satisfying counting book.

Moss, Lloyd. *Zin! Zin! Zin!: A Violin*. Illustrated by Marjorie Priceman. New York: Simon & Schuster Books for Young Readers, 1995.

ISBN 0-671-88239-2 $15.00

★★ Single concept, PS–2

One by one musicians playing ten different instruments enter the concert floor. The book's main purpose is to introduce musical instruments, but it cleverly combines the names for different-sized groups by introducing each group from one through ten. As each sequential number is introduced, each instrument is strikingly described in the poetic text. "With mournful moan and silken tone, / itself alone comes ONE TROMBONE. / Gliding, sliding, high notes go low; / ONE TROMBONE is playing SOLO." A duo forms when the trumpet player arrives, and a trio is seen when the French horn player joins them onstage. Appropriate number names for each musical group are included, such as *nonet* for nine musicians; a chamber group of ten appears for the final group. Classification is also included, since a final page shows the players grouped by string, reed, or brass instruments The watercolor illustrations are playful and flowing and create pleasing scenes; the book was noted as *New York Times* Best Illustrated in 1995.

Murphy, Stuart J. *Every Buddy Counts.* Illustrated by Fiona Dunbar. New York: Harper Collins Publishers, 1997.

See Series and Other Resources: MathStart

O'Brien, Mary. *Counting Sheep to Sleep.* Illustrated by Bobette McCarthy. Boston: Little, Brown & Co., 1992.
ISBN 0-316-62206-0 $13.95
☆ Single concept, PS–1

Pastel illustrations show a farmer who is cheerfully checking her animals in the diminishing daylight. Numbers from one through four are repeated in verse throughout this bedtime story: "Drowsy hens in henhouse gather. / One cluck, / two cluck, / three cluck, four." The peaceful scenes are interrupted by four rambunctious sheep who create chaos as they mischievously romp from one situation to another. The farmer finally settles them for the night; three sleep by the barn and the fourth cuddles near the sleeping farmer. Word names for numbers are included throughout the rhymes; often the animals are to be counted and the sound that the animals make is repeated an appropriate number of times in the verse.

Ochs, Carol Partridge. *When I'm Alone.* Illustrated by Vicki Jo Redenbaugh. Minneapolis, Minn.: Carolrhoda Books, 1993.
ISBN 0-87614-752-X $18.95
ISBN 0-87614-620-5   (paper)   6.95
★★ Single concept, PS–3

The beautiful dreamlike quality of the colored-pencil artwork is perfect to accompany this charming tale of a little girl's imagination. The detailed representation of her bedroom as she wakes provides the foreshadowing of what is to come. Her stuffed animals become the stimulus for a wild imagination as ten aardvarks make a mess of her food, nine lions untidy her bed, eight turtles mess up the bathroom, and seven camels tangle the garden hose. The rhyming text captures the little girl's indignation as she explains away to herself the messes created by her visitors. "I saw ten aardvarks in my food, / And their behavior was *awfully* rude. / It put me in a grumpy mood." The related scene shows the little girl stretching to reach a box on a high kitchen shelf while three aardvarks are seen knocking objects off the shelves, one peeks out of the open cupboard below, three more are near the open refrigerator door, two are under the

stove, and another one opens a stove door. This is typical of the
appealing tasks for counting sets from ten to one. The text and
illustrations combine for a pleasant, satisfying experience of getting
to know a charming and imaginative character. Mom is included at
times as someone who just came walking out of her study to inquire
about the situation. Both characters are African American.

Piers, Helen. *Is There Room on the Bus?* Illustrated by Hannah Giffard.
New York: Simon & Schuster Books for Young Readers, 1996.
ISBN 0-689-80610-8 $14.00
★★ Single concept, K–2

Sam fixes up his battered, rickety yellow bus in preparation for a
trip around the world. In South Africa he meets one lonely lion who
asks to join him. Because there is "plenty of room on the bus," new
fellow travelers are added at each stop. Sam and the one lonely lion
are joined by "two cross cows, three wet walruses, four enormous
elephants, ..., nine crotchety crocodiles" as the bus bulges at the
seams. When ten bothersome bees buzz their way onto the bus, havoc
erupts; the animals burst out of the bus to escape those bothersome
bees. As each new group from one through ten is encountered, it is
shown on a separate page for ease in counting. On the opposite page
the new group is shown with *all* the bus's occupants; the text lists the
accumulated groups with the number words embedded within the
text in bold print. Numerals are included on the well-designed
summary pages, which rename and show each group of animals. The
animals offer to go on the next trip with Sam, but "Luckily the ten
bothersome bees flew away." A final challenge is to find and count all
ten bees. Vibrantly colored illustrations complement this humorous
story. Note: The story takes an unusual route around the world, as
Sam and his friends travel from South Africa to Ireland to Canada to
Burma and later travel back to India and Egypt.

Powers, Christine. *My Day with Numbers.* New York: Scholastic, 1992.
ISBN 0-590-44975-3 $4.95
☆ Single concept, PS–K

Simple text and watercolor illustrations depict a young child's
experiences with counting throughout her day. From "one child / two
slippers / three muffins" to the final ten yawns, the numbers from
one through ten are clearly depicted in this cardboard book. The
numerals from 1 through 10 are shown next to ten small illustrations
of a yawning little girl, with a final good-night on the closing page.

Reiser, Lynn. *Christmas Counting.* New York: Greenwillow Books, 1992.
ISBN 0-688-10676-5 $14.00
ISBN 0-688-10677-3   (library binding)   13.93
★★★ Single concept, PS–4 (PS–3)

Much more than a counting book, this story centers on a growing family whose members establish their own traditions as they celebrate the joy of Christmas from year to year. On the first Christmas, a young man builds a house in the clearing in the forest, brings a tiny tree into the house, and puts 1 star on top of it. After Christmas he lovingly replants the tree. Subsequent Christmases show him with his expanding family: a bride, a puppy, a child, a cat, a new baby, a parrot, a pony. Each year he continues the tradition of digging up the growing tree. Each year new decorations are placed on the tree, with the new family member adding his or her distinct ornaments. In the second year the wife adds 2 hearts, in the third year the puppy adds 3 paper chains, in the fourth year the baby hangs 4 angels, and so on. By the eighth year the tree needs to be dragged in by the pony; by the ninth year the tree is too big to be moved, but the tradition is continued with another tree. The original tree is decorated with 9 sets of lights while the family sings 10 carols to it. Through both text and illustration, the accumulative contribution to the tree of each "family" member is depicted: "put 1 star on top of it, / the mother put 2 hearts on it. / The dog put 3 paper chains on it." This provides another opportunity to recount the accumulating decorations and to observe the changes in the family members. The celebration of the joy of Christmas and of the warmth of a loving family illustrates the passage of time and the changes it brings in plants, animals, and humans. The accumulating counting opportunity for young readers is merely a byproduct of a well-told, heart-warming story. The text is interspersed with simple, colorful illustrations completed with watercolor and black pen.

Ricklen, Neil. *My Numbers* • *Mis Números*. New York: Aladdin, 1994.
ISBN 0-689-71770-9 $3.95
★★ Single concept, PS–K

From the cover to the final page, winsome toddlers grin and giggle; their personalities project from the glossy full-color photographs. Twelve beautiful children offer the reader items to count: one guitar, two flowers, three blankets, four presents, ..., ten hats. Each page provides numeral, number word, and name of object first in English and then in Spanish. Underneath both the English and the Spanish text is a pronunciation guide. This sturdy, little board book will withstand the expected rough treatment of a toddler.

Ruschak, Lynette. *The Counting Zoo: A Pop-up Number Book.*
Illustrated by May Rousseau. New York: Aladdin Books, Macmillan Publishing, 1992.
ISBN 0-689-71619-2 $13.95
★★★ Single concept, PS–2

Excellent design is the hallmark of this unique counting book. As the reader turns each page, a large numeral unfolds and extends to a height almost double the height of the page. The objects to be counted are stuffed animals. The pig on the first page has a yellow numeral 1 emblazoned on the front of its clothing. The large numeral is the same bold yellow. The colors of the numerals and animals are in contrast to the stark white background. The only text is the word name printed in gray. Each previously introduced animal can be found frolicking on the subsequent pages. Such surprises as finding the flamingo sitting on top of the numeral 7 are a delight. The young reader will look for and count in order each winsome animal on each page. The closing back cover provides a delightful ending. Ten square doors with the numerals from 1 through 10 are arranged in two rows; the color scheme matches the fold-out numerals. As each door is opened, the reader sees a familiar character pop out, with the pig waving one hat, the flamingo holding two watches, the hippopotamus surrounded by seven presents. And on the final flap are the word names for one through ten.

Scott, Ann Herbert. *One Good Horse: A Cowpuncher's Counting Book.* Illustrated by Lynn Sweat. New York: Greenwillow Books, 1990.

ISBN 0-688-09146-6 $12.88
ISBN 0-688-09147-4   (library binding)   12.95

★★ Single concept, PS+

Pencil and transparent oil paints are masterfully used to create dawn and early morning light changes in a western landscape. The calm and beauty of the scenery are reinforced by the constant rhythm of an add-on journey as a little boy and his father ride one good horse to check their cattle. The two buckaroos see three cow dogs and four old bulls before full sun. After seeing ten mountain quail, they check fifty cedar fence posts before they finally spot 100 head of cattle grazing in the meadow. Opposite each illustration is the simple text on a page framed with an orange and turquoise southwestern design; only number word names are included in the text. The final scene provides a panoramic view of the entire journey with the counted objects in the distance as father and son ride off the right-hand side of the page—a satisfying ending to a simple, but beautiful book.

Serfozo, Mary. *Who Wants One?* Illustrated by Keiko Narahashi. New York: Margaret K. McElderry, 1989.

ISBN 0-689-50474-8 $13.95
ISBN 0-689-71642-7   4.95

☆ Single concept, PS–3

"Do you want ten? Ten speckled eggs from ten brown hens; ten pink pigs happy in their pens." With a wave of a magic wand, this vision appears when a young girl inquires, "Do you want ten?" Her younger brother stubbornly responds to each proposal, "No, I want one!" Each number is offered. Each offer is rejected. Fanciful watercolor illustrations depict more than one set to count on every page. Often the objects are shown in one-to-one correspondence, such as four goats in four boats. Counting challenges appear throughout the book; in the set of ten eggs, one has broken and a baby chick is emerging; in a set of eight plates, one is shattered and another split; one scene of seven shows two adult geese and five goslings. The little boy is adamant in saying, "No, I want one." At last the reader understands as the boy is handed *one* puppy. The end pages summarize the numerals, number names, and sets from one through ten.

Sugita, Yutaka. *Goodnight, 1, 2, 3.* New York: Scroll Press, 1971.
ISBN 0-87592-022-5 $9.95
★★★ Single concept, PS–2

Vibrant colors glow from each rich, magnificent page of this bedtime counting book. A sleepless child sequentially counts one through eleven colorful, appealing animals and objects. This beginning counting book has no text; only the numeral, the appropriate number of objects, and the exploring child appear on each page. Beginning place-value concepts are depicted on the pages representing 11; ten balloons are on the left-hand page, and the eleventh balloon is tied to the 1 in the ones place in the numeral 11 on the right-hand page. Although this is a single-concept book, it contains opportunities to teach color and spatial relationships.

Tafuri, Nancy. *Who's Counting?* New York: Greenwillow Books, 1986.
ISBN 0-688-06130-3 $15.00
ISBN 0-688-06131-1   (library binding)   14.93
☆ Single concept, PS–1

Puppy sets out to explore the environment and finds "1 squirrel, 2 birds, 3 moles, ..., and 10 puppies." The last page shows ten puppies in one-to-one correspondence with ten food dishes labeled with the numerals from 1 through 10. This simple text accompanies attractive drawings of the appropriate number concept. Only numerals are introduced in the text.

Time-Life for Children Staff. *How Many Hippos? A Mix-&-Match Counting Book.* Illustrated by Robert Cremins. Alexandria, Va.: Time-Life Incorporated, 1990.

See Early Number Concepts: Counting—Other Number Concepts

Tryon, Leslie. *One Gaping Wide-Mouthed Hopping Frog.* New York: Atheneum, 1993.
ISBN 0-689-31785-9 $14.95
★★ Single concept, PS–1

Let's see if we can keep pace with the Pleasant Valley mail carrier, a wide-mouthed hopping frog, as he delivers the mail to the residents of his busy little town. On his route, we'll catch sight of "2 birthday cakes for a very old dog; 3 monkeys dancing the clog; 4 ostriches who like to jog" and many more delightful surprises. After the numbers from 1 through 5 are introduced, an illustration shows the 5 puppies, 4 ostriches, 3 monkeys, and 2 birthday cakes. All the captions are repeated but in reverse order, from 5 through 1. The rhythmic text and detailed illustrations make an exhausting but enjoyable trip as the delivery continues from 6 beetles through 10 peacocks. The illustrations are made to pore over and return to many times, or a reader might miss Gracie's cheating ways at the Magic Apple stand, or the pup who stole a lobster, or the sky-diving peacocks, or the irony of beavers representing joiners, or the puppy tugging at Ms. Golden's coat. There is much intricate detail in the charming, whimsical and attention-grabbing illustrations. Views of all the colorful characters from 1 through 10 can be found in a final busy scene; a final count shows each group organized and named by sets from 10 through 1. The satisfaction from, and a reward for, a job well done can be appreciated when the hopping frog is viewed in his den with hot tea and a baked bug—and a bucket of water to soak his feet.

Tucker, Sian. *Numbers.* New York: Little Simon, 1992.

ISBN 0-671-76908-1 $2.95

☆ Single concept, PS–K

Bold color dominates the backgrounds and the objects to be counted in this tiny cardboard-paged book. Minimal text includes the numeral and the names for the numbers and the objects. The appropriate number of dot counters appears above each numeral. On the opposite page is another opportunity to count, such as three chairs, six fish, or nine apples. After the numbers from one through ten are represented, twenty airplanes are pictured in a 4-by-5 array. The last page is simply "lots of spots."

Turtle Count. Illustrated by Norman Gorbaty. New York: Little Simon, 1991.

ISBN 0-671-74434-8 $3.95

☆ Single concept, PS

Numbers from one through five are the focus of this small, sturdy board book. Five different scenes introduce the inquisitive turtles, with one more in each scene. A large bold black numeral, the word name, and easy-to-count turtles in uncluttered illustrations are introduced by the simple rhyming text designed for preschoolers. An appealing aspect of the book is its turtle shape.

van der Meer, Ron, and **Atie van der Meer**. *Funny Hats: A Lift-the-Flap Counting Book with a Surprise Gift.* New York: Random House, 1992.

ISBN 0-679-82850-8 $7.99

☆ Single concept, PS

Hiding behind eleven different, hat-shaped flaps are countable collections from one clever cricket, two not-so-happy hippos, …, to ten silly spiders and twenty cuddly cats. As each hat is lifted, the members of the set appear in miniature hats similar to the flap hat. Each set is announced in simple text and is often humorously posed, such as the "nine magnificent mice" standing in a chorus line. Readers may guess some of the hidden sets, since a glimpse of animals appears beyond the outline of the hat on some pages. The numbers from one through ten and the number twenty are represented; a large black numeral appears in the upper corner of each page. The end pages contain a summary of all the numbers, with the appropriate numeral printed beside the appropriate number of miniature hats.

Van Fleet, Matthew. *One Yellow Lion.* New York: Dial Books for Young Readers, 1992.

ISBN 0-8037-1099-2 $7.95

★★★ Single concept, PS+

Clever design is a strong feature in this durable, little board book with foldout pages, which features the numbers from one through ten, as well as the names of ten colors and ten animals. A large yellow numeral 1 is on the opposite page from the words "One / Yellow" printed in large type. As the page with the numeral is folded back, a simple illustration of a friendly yellow lion appears, along with the word *Lion.* Part of the illustration of the yellow numeral 1 is now seen as part of the lion's tail. This unusual design, incorporates part of the colored numeral into the illustration of the featured animals, will challenge the reader to predict what is to come. For another surprise, the last five pages unfold to display all ten sets of animals in the ocean or on the beach. The reader will search to count sets like the five blue whales, the six black cats, or the eight brown bears. On the flip side of these pages is a four-page-high illustration of most of the animals as they form a very precarious pyramid.

Voss, Gisela. *Museum Numbers.* Illustrated by Kyung Mi Ahn. Boston: Museum of Fine Arts, 1993.

ISBN 0-87846-370-4 $14.00

★★ Single concept, PS–1 (1–3)

Paintings by John Singer Sargent and Vincent Van Gogh, mummies from Egypt, and an embroidered silk robe from Japan are some of the art objects from the Boston Museum of Fine Arts that are reproduced in this counting book. Opposite each art reproduction is a page designed in two colors that announces the number to be counted. Sargent's *The Daughters of Edward D. Bolt* is opposite a blue and white page stating "4 Girls" above silhouettes of four girls. The *Furisode,* an embroidered silk robe from the Edo period of Japan, is opposite an orange and yellow page stating "9 Birds" above silhouettes of nine cranes. The appropriate numeral is repeatedly printed along the right-hand border of each number page. Because the cardboard pages are layered in increasing size, each colorful border of the numerals from 1 through 10 provides an attractive view of the closed book. Bibliographic entries for each featured art object are included in the end pages, as well as an invitation to visit the Boston Museum of Fine Arts or galleries or museums near the reader. The clever design of this counting book makes it an attractive introduction to counting and the arts.

Wegman, William. *One, Two, Three.* New York: Hyperion Books for Children, 1995.

ISBN 0-7868-0103-4 $6.95

☆ Single concept, (PS–1)

Wegman's famous photographs of weimaraners effectively illustrate the numbers from 1 through 10 in this simple, no-text counting book. Each square cardboard page introduces the number concept and its corresponding numeral; no word names are included. Black-and-white photographs depict weimaraners carefully arranged to form each numeral; 4 dogs curved in a sleeping position form the numeral 3. The color photograph below shows 3 brown dogs looking up at the camera while stretching out on a tan floor. Some illustrations are humorously done, such as the 2 dogs sitting attentively in schoolroom desks or the 4 dogs standing on their hind legs while dressed in hooded bathrobes. The final page shows 10 appealing puppies sleeping in a row.

Wells, Rosemary. *Max's Toys: A Counting Book.* New York: Dial Books for Young Readers, 1979.

ISBN 0-8037-6068-X $4.50

★★ Single concept, PS–K

Max and his sister Ruby are involved in a debate on toys in this little, sturdy, cardboard counting book. Max would like Ruby's doll, Emily, but Ruby announces, "Emily is 1 thing you may not have." Max considers his two toy soldiers, a house with three windows, four stuffed animals, ..., and a ten-car train and trades all his toys for Emily. The simple colorful pictures of the two rabbits, Max and Ruby, are charming. The objects are easy to count, although more than one set is introduced on most pages. This simple counting book provides an interesting introduction to Max and Ruby, who appear in other books by Wells.

Wolff, Ferida, and **Dolores Kozielski.** *On Halloween Night.* Illustrated by Dolores Avendaño. New York: Tambourine Books, 1994.

ISBN 0-688-12972-2 $15.00
ISBN 0-688-12973-0   (library binding)   14.93

☆ Single concept, PS–1

One witch, two black cats, three owls, ..., eleven bats, and thirteen ghosts are some of the familiar Halloween sets to be counted. The attractive illustrations effectively create the excitement of Halloween. The eerie scenes painted in dark shades of browns, greens, and blues sequentially represent the numbers from one through an unlucky thirteen. Some sets, such as the nine crows, eleven bats, or twelve ghosts, are more challenging to count because of overlapping and shadowy forms. Parallel, poetic text complements each scene: "Three owls hunt / through the forest mist, / grabbing this, grabbing that, / hooting at each startled shrew, / on Halloween nights. / Oo-oo-oo-ooh.'

Wood, Jakki. *Moo Moo, Brown Cow.* Illustrated by Rog Bonner. San Diego, Calif.: Gulliver Books, Harcourt Brace & Co., 1992.

ISBN 0-15-200533-1 $12.95

★★ Single concept, PS–2

Integrated into this counting book are the names of different colors and the names of selected adult animals and their offspring. "Honk honk, blue goose, / have you any goslings?" is an example of the inquiries a lone kitten makes to the adult member of each new group of animals. "Yes kitty, yes kitty, / five fat goslings" is the reply as five goslings snuggle around the blue goose for the reader to count. Parallel text is used throughout the book as the kitten encounters a brown cow and its calf, a yellow goat and its kids, an orange hen and her chicks, and others. The sets of baby animals are ordered in a sequence from one through ten. The concept of zero is alluded to with "No kittens, no kittens, / but many, many friends." Only word names for the numbers are introduced and they are embedded in the text. One color dominates each of the pleasing watercolor illustrations.

Woodard, James, and **Linda Purdy.** *One to Ten and Count Again.* Illustrated by James Woodard. La Puente, Calif.: Jay Alden Publishers, 1972.

ISBN 0-914844-07-5 $6.89

☆ Single concept, PS–K

Black-and-white photographs illustrate this simple counting book. Most of the pictures depict children happily engaged in play activities; other pictures represent objects to be counted, such as baby chicks, balloons, or toys. A contrived rhyming text introduces numerals and number names, for example, "4, four, four, toys on the floor."

Chants, Songs, and Rhymes

Traditional songs and rhymes have withstood the test of time. Their rhythm and the predictability and repetition of their action and rhyme have made them favorites of children through the ages. These very characteristics make them a suitable resource for combining beginning reading and mathematics experiences. As children sing along with the countdown of Gerstein's version of *Roll Over*, they use the predictability to help them track print. The familiar melody found in Brett's *The Twelve Days of Christmas* is revisited in Manushkin's *My Christmas Safari* and O'Donnell and Schmidt's *The Twelve Days of Summer*. Different adaptations of familiar rhymes are found in this section, as well as new rhymes and poems.

Arnold, Tedd. *Five Ugly Monsters.* New York: Scholastic, 1995.
ISBN 0-590-22226-0 $6.95
☆ Single concept, PS–K

Counting back, or "one less than" is the mathematical focus of this sturdy book. The story line is a unique adaptation of a familiar chant: "Five ugly monsters jumping on the bed. One fell off and bumped his head." One by one, monsters fall off the bed; each time the young boy telephones the sleepy doctor who prescribes no more jumping on the bed. The monsters are ugly, friendly looking globs who scramble in and out of the child's bedroom with a different, although appropriately smaller, group for each countdown. A surprising twist to this version occurs when the doctor cuts the telephone cord during the last call. The young boy is forced to be the ultimate master, roaring the final and strongest demand for "NO MORE MONSTERS JUMPING ON THE BED!" to all the monsters who have sneaked back into the room. The cartoonlike illustrations are colorful and printed on card stock.

Baker, Keith. *Big Fat Hen*. San Diego, Calif.: Harcourt Brace & Co., 1994.

ISBN 0-15-292869-3 $13.95

★★ Single concept, PS–1

"1, 2, ... / buckle my shoe / 3, 4, ... / shut the door"; this familiar nursery rhyme is complemented by Baker's illustrations and interpretations. Large colorful illustrations spread across each double page. On the first page, with the text of "1, 2, ...," a colorful hen with a bright red comb and feathers of gold, green, and blue chases one dragonfly while two eggs sit in the straw. The next page shows 2 newly hatched chicks exploring the buckles on a shoe. The pattern continues as new hens with feathers of varying shades of blue, yellow, and fuchsia chase away an odd number of insects while their even number of eggs wait to be hatched. Distinctly designed and beautifully shaded in rich colors, a total of 6 "fat hens" populate the scenes. The "big fat hen" and her five friends are lined up and readily counted, but the challenge is to count "all their eggs ... and all their chicks!" Although sets of two, four, six, eight, and ten eggs wait to be counted throughout the book, these final pages depict a total of fifty eggs and forty-nine chicks. It may appear that a chick is missing, but one chick can be found on the endpapers, chasing a dragonfly. Baker's *Big Fat Hen* received the Golden Kite Award from the Society of Children's Bookwriters and Illustrators.

Beck, Ian. *Five Little Ducks*. New York: Henry Holt & Co., 1992.

See Early Number Concepts: Counting—Operations

Brett, Jan. *The Twelve Days of Christmas*. New York: Putnam, 1990.

ISBN 0-399-22197-2 $14.95
ISBN 0-399-22037-2   3.95

★★ Single concept, all ages

Brett's intricately illustrated scenes with elaborate borders dramatically and beautifully portray the action and content of this traditional Christmas carol. As each consecutive verse is introduced, the new gift is shown in rich colors and detailed design. Six geese a-laying are shown in fanciful headdresses and vests, and another opportunity to count to six is in the top border, which shows six Christmas ornaments in the shape of geese hanging from a Christmas tree. Each double-page spread has side panels illustrating a family preparing for the holidays. The family scenes reflect traditions from different nationalities and include greetings in different languages.

Bursik, Rose. *Zoë's Sheep*. New York: Henry Holt & Co., 1994.

See Early Number Concepts: Counting—Operations

Christelow, Eileen. *Five Little Monkeys Jumping on the Bed.* New York: Clarion Books, 1993.

ISBN 0-395-55701-1   (paper) $5.95

☆ Single concept, (PS–K)

"One fell off and bumped his head. The mama called the doctor...." This familiar chant illustrated in pastels should delight young readers. The repetitive chant and humorous scenes show the monkey siblings jumping on the bed, the inevitable falls, the calls to the doctor, and the mama monkey caring for a crying little monkey and repeating the doctor's advice. The final pages show all five little monkeys safely tucked in bed and mama monkey retiring to her room to jump on her bed! The story line also introduces beginning subtraction concepts as one by one, the number of monkeys jumping on the bed diminishes.

———. *Five Little Monkeys Sitting in a Tree.* New York: Clarion Books, 1991.

ISBN 0-395-66413-6   (paper) $5.95

☆ Single concept, PS–K

This format and pastel illustrations are similar to those in *Five Little Monkeys Jumping on the Bed;* a repetitive chant and humorous scenes show the five monkey siblings disappearing one by one. The monkeys are sitting in a tree and teasing Mr. Crocodile by daring him, "Can't catch me!" Each challenge is followed by a snap from Mr. Crocodile's jaws, and one fewer monkey is in the next scene. The monkey's mama and other spectators look on with dismay, but in the final scenes, they find all five monkey siblings hiding in the tree. This story line offers a simple introduction to the concept of subtraction.

Chwast, Seymour. *The Twelve Circus Rings.* San Diego: Gulliver Books, Harcourt Brace Jovanovich, 1993.

See Early Number Concepts: Counting—Operations

Clarke, Gus. *Ten Green Monsters.* New York: Golden Books, 1993.

ISBN 0-307-17605-3 $14.95

★★ Single concept, PS–3

"Ten green monsters standing on the wall, ..., And if one green monster should accidentally fall..." is the familiar chant from Clarke's delightful "Lift-the-Flap-and-See-Them-Fall Book." Ten friendly monsters are observed standing in line along a brick wall; a small green hand indicates where the large page flap is to be turned. The next line of the chant, "there'll be nine green monsters standing on the wall," is uncovered, and nine monsters are now pictured on the wall with one falling monster exclaiming, "AAAARGH!" The pattern continues as one by one they tumble, slip, or slide off the wall. Lively banter among the fallen monsters is recorded in a "balloon" format; their conversations and antics add humor to the text. Gradually they question whether someone had pushed them. When one lone monster is left, the others climb back onto the wall and this one lone monster does *not* accidentally fall. Throughout the pages the reader notices that this same monster changes position from one section of the wall to the opposite and is always standing next to the monster that "accidentally falls." This well-designed sturdy book with its appealing cast of cartoon characters should be a favorite with young children and lead naturally to beginning subtraction concepts.

Coats, Laura Jane. *Ten Little Animals.* New York: Macmillan, 1990.
ISBN 0-02-719054-4 $11.95
☆ Single concept, PS–1

Rich pencil line drawings and fresh watercolors recreate the world, a boy, and his ten stuffed toys as they jump on the bed. One by one, each toy bounces off and bumps its head. From bed to toy box, the animals are systematically transferred after a call on the toy phone to ask the doctor's advice. The doctor's final admonition, "Tell those animals it's time to go to bed!" is a satisfying conclusion as the little boy kisses each little, sleepy animal on the cheek and climbs wearily into bed himself. The charming illustrations on the left-hand page are bordered in a primary color, which helps contain the havoc created by the jumping friends. A pictorial clue is provided to help children predict who will fall next, since one animal's body always breaks through that border. Printed above the rhyming text, a large numeral, the same color as the opposite page's border, announces the number of animals still jumping on the bed. The countdown focus of the story is reversed on the last page in a pictograph showing the one to ten animals with the numerals and number names for each set.

Dale, Penny. *Ten out of Bed.* Illustrated by Frances Lloyd. Cambridge, Mass.: Candlewick Press, 1993.
See Early Number Concepts: Counting—Operations

Dunrea, Olivier. *Deep Down Underground.* New York: Macmillan Books for Young Readers, 1989.
ISBN 0-02-732861-9 $14.95
ISBN 0-689-71756-3   (paper)   4.95
☆ Single concept, PS–2

"4 furry caterpillars scooch and scrunch when they hear; 3 big black beetles scurry and scamper when they hear; 2 pink earthworms wriggle and wrangle when they hear; 1 wee moudiewort digging and digging, deep down underground." From one through ten, accumulating sets of underground animals in their habitats are effectively illustrated by pen and ink with watercolor. Alliterative text introduces each new group and repeats the refrains from the previous pages. The young reader should delight in the text as well as in finding and re-counting the previously named underground creatures.

Gerstein, Mordicai. *Roll Over!* New York: Crown Publishers, 1984.
ISBN 0-517-55209-4 $12.00
★★★ Single concept, PS–1

This beautifully illustrated counting book uses fold-out flaps to reveal the consequences of rolling over in a crowded bed with ten inhabitants. Readers can soon predict the outcome of the small boy's command, "Roll over," but the uniqueness of each animal who falls out of bed still provides an element of surprise. Repetition makes this book easy for beginning readers; the book also depicts subtraction as counting back by showing one fewer in the bed. Even the passage of time is implied by the ever-changing view through the night window.

Goennel, Heidi. *Odds and Evens: A Numbers Book.* New York: Tambourine Books, 1994.
ISBN 0-688-12918-8 $15.00
ISBN 0-688-12919-6 (library binding) 14.93
☆ Single concept, PS–1

Thirteen familiar expressions, such as "a one-horse town," "two in the bush," and "three blind mice," sequentially introduce the numbers from one through thirteen in this number book. Each phrase is framed by the numerals 1 through 13, which form a top and bottom border; on a single-color background, the numerals are printed in white, except the number being featured on the page. What is to be counted varies in complexity. The page opposite the phrase "the five senses" shows a woman's face with her hand against her cheek. "Seventh heaven" features an illustration of an angel among seven clouds. "Behind the eight ball" features the eight ball on a pool table; some of the other seven billiard balls are partially shown. The illustrations are composed of bold colors with minimum detail. Although *Odds and Evens* is the title of the book, that concept is not developed or referred to in the book.

Katz, Michael Jay, ed. *Ten Potatoes in a Pot: And Other Counting Rhymes.* Illustrated by June Otani. New York: Harper & Row, 1990.
ISBN 0-06-023107-6 (library binding) $12.98
ISBN 0-06-023106-8
☆ Single concept, PS–2

"One, two—buckle my shoe / Three, four—shut the door" are the familiar lines that begin this counting book of rhymes. The introductory poem features the counting numbers from one through twenty; the closing poem highlights very large numbers. Both familiar and not-so-familiar counting rhymes have been selected and are organized by the numbers from one through twelve. A black-and-white border across the top of the page highlights each numeral and silhouettes the appropriate number of objects to be counted. A simple, black rectangular border sets off each poem. Charming pastel illustrations, reflecting the poem's content, provide another opportunity for counting.

Langstaff, John. *Over in the Meadow.* Illustrated by Feodor Rojankovsky. New York: Harcourt, Brace & Co., 1992.

ISBN 0-15-258853-1 $19.95

☆ Single concept, PS–1

Colorful pastels alternate with soft brown drawings to capture nature's creatures in the beautiful meadow. Each rich scene shows a mother and her young in their natural habitat, from old mother turtle and her little turtle to mother rabbit and her ten bunnies. Discriminating which objects to count will challenge the young reader in some scenes. The book closes with the text and the music for the song "Over in the Meadow."

Loveless, Liz. *One, Two, Buckle My Shoe.* New York: Hyperion Books for Children, 1993.

ISBN 1-56282-477-5 $13.95
ISBN 1-56282-478-3   (library binding)   13.89

☆ Single concept, PS

Rote counting from one through twenty is encouraged in this familiar rhyme. The treatment is unusual because of the artist's style. Loveless has chosen to represent up-to-date characters for this traditional rhyme, which usually evokes images of a historical time setting. For "17, 18, Maids a-waiting," the pastel illustration depicts a little girl eating an ice-cream cone while waiting for a bus. On each left-hand page, block print is used to feature large numerals and the verse on a pastel background. The same pastel color is used for the background for the dreamlike illustrations on the opposite page. The unique illustrations, reminiscent of those of Toulouse-Lautrec with minimal detail and strong outlines, are appealing. The function of the artwork is to illustrate the verse, not the number concepts.

Mack, Stan. *Ten Bears in My Bed.* New York: Random House, 1974.

ISBN 0-394-92902-0   (library binding) $11.99

☆ Single concept, PS–1

Based on the familiar countdown rhyme, Mack's adaptation is imaginative and comical. In response to the demand to roll over, each bear chooses to leave in a different way; one flies out, one gallops out, one skates out, one roars out, …, until the last one rumbles out. The illustrations are lively and humorous and totally support the actions described in the text. The one-by-one decreasing size of the set lends itself to "one less than" and beginning subtraction concepts. Although this is a good book, better adaptations are available.

Manushkin, Fran. *My Christmas Safari.* Illustrated by R. W. Alley. New York: Dial Books for Young Readers, 1993.

ISBN 0-8037-1294-4 $ 13.99

ISBN 0-8037-1295-2   (library binding)

★★ Single concept, PS–1 (PS–2)

Action-packed and detailed, the hilarious illustrations capture the excitement of this unique rendition of the familiar cumulative Christmas song, "The Twelve Days of Christmas." A young girl opens her Christmas presents to find a set of African animals; the subsequent tale is her dream of an imaginary photographic safari with her father. "On the first day of Christmas, ..., a green truck for our safari, on the second day ... two leopard cubs and a green truck for our safari." On each double page the fascinating watercolor illustrations become more and more crowded with three wildebeests, four shy giraffes, ..., and finally ten elephants. The young reader will be absorbed in looking for all the animals in the sets, not just to count them, but also to check on what they are up to! Baboons steal a camera, leopards pounce on one another and hide in the grass, while giraffes nose into the back of the green truck. The back cover of the book reprints the music for the song.

Marzollo, Jean. *Ten Cats Have Hats: A Counting Book.* Illustrated by David McPhail. New York: Scholastic, 1994.

ISBN 0-590-20656-7 $6.95

☆ Single concept, PS–K

Repetitive and rhyming text introduces the numbers from one through ten: "One bear has a chair, / but I have a hat. / Two ducks have trucks, / but I have a hat." One child, pictured with a different, but appropriate hat in each scene, is the narrator. Wearing a farmer's hat, the child visits five pigs with wigs; a miner's hat complete with light is donned to view nine snails with trails. The final surprise is the set of ten cats, each with a hat just like the hat of the child, who is seen on the last page with the ten hats. An appropriate numeral is prominently displayed in the upper left-hand corner of each scene. Pastel illustrations complement the text with charm and humor.

Merriam, Eve. *Train Leaves the Station.* Illustrated by Dale Gottlieb. New York: Bill Martins Books, 1994.

ISBN 0-8050-3547-8   (paper) $5.95

☆ Single concept, PS–2

"Hunter on the horse, fox on the run, train leaves the station at one-o-one." These are the opening lines of a poem, originally published in *You Be Good & I'll Be Night,* that is repeated as the text in this simple book. A large numeral 1 appears with the first verse. The circular front of the train's engine is a clock face whose hands depict 1:01. Numerals from 2 through 10 are introduced similarly. No objects to represent the numbers until the end pages. On this double-page spread, numerals from 1 through 10 are shown with the appropriate number of counters from one fox to ten cows. Gottlieb's artwork is simple and childlike in style. The few details are outlined in black and dominated by strong color.

Nikola-Lisa, W. *No Babies Asleep.* Illustrated by Peter W. Palagonia. New York: Atheneum Books for Young Readers, 1994.

ISBN 0-689-31841-3 $15.00

Single concept, PS–1

Although this may be another adaptation of that familiar favorite, *Ten in the Bed,* the story line is a little discomforting. The rhythmic and repetitive text is charming, but the action of the story is somewhat threatening as various groups of zoo animals climb through the nursery window to take yet another sleeping baby from the crib. The pastel illustrations of monkeys, turtles, peacocks, koalas, seals, and so on, depict the action 'Take one out, / And pass it about" while the remaining babies lie innocently asleep. One by one, each of the ten babies disappears through the window. The repetition and predictability of the text is broken when the same animal takes baby two and baby one. The sinister, foreboding atmosphere continues as a woman views the empty nursery and then searches for the missing children. The babies are found being entertained at the zoo by the animals. Recognizing children's fear of separation, a major fear for them, we would not share this book with young children. It is possible that the same story line could be made more palatable with illustrations that clearly depict the excitement and joy of each awakened baby.

O'Donnell, Elizabeth Lee, and **Karen Lee Schmidt.** *The Twelve Days of Summer.* Illustrated by Karen Lee Schmidt. New York: Morrow Junior Books, 1991.

ISBN 0-688-08202-5 $15.00
ISBN 0-688-08203-3   (library binding)   14.93

☆ Single concept, PS–3

One hardly expects that while looking at sandy scenes beside the sea, the "Twelve Days of Christmas" melody would be running through one's head. But from the opening line, "On the first day of summer, / I saw down by the sea / A little purple sea anemone," to the final line, one hears the familiar melody. "Eight crabs a-scuttling, /

Seven starfish strutting, / Six squid a-swimming" are some of the creatures that a little girl encounters on her explorations on twelve summer days. Each scene becomes more and more crowded as the creatures accumulate; the reader's challenge is to count each new set as well as to find and count the previously introduced sets. Charming illustrations in watercolors complement the verse.

O'Keefe, Susan Heyboer. *One Hungry Monster: A Counting Book in Rhyme.* Illustrated by Lynn Munsinger. Boston: Little, Brown & Co., 1992.

ISBN 0-316-63385-2	$12.95
ISBN 0-316-63388-7 (paper)	5.95

★★ Single concept, PS–3

Rollicking rhyme, an outrageous story line, and action-packed illustrations make this a story that demands to be repeated time and again. One by one, ten hungry monsters demand food from the formerly sleeping little boy. Each growing set of monsters causes more havoc in its voracious search for food: "Eight hungry monsters / on the chandeliers, / swear they haven't eaten / for maybe twenty years." A new opportunity to count to ten follows as the little boy scrambles to serve one jug of apple juice, two loaves of bread, ..., ten jars of peanut butter. The first counting activities involve word names, whereas the second section uses numerals. The story line continues as the monsters devour the food and cause more messy mania in the house until the little boy demands that they leave. The pastel illustrations, reminiscent of those in *Where the Wild Things Are,* capture the hilarious antics of the naughty, but lovable monsters. Young children will delight at the possibilities for future adventures suggested by the final illustrations in the book.

Peek, Merle. *Roll Over!* New York: Clarion Books, 1981.

ISBN 0-395-29438-X	$14.95
ISBN 0-395-58105-2	5.95

☆ Single concept, PS–2

Nine animals, including a raccoon, a parrot, a bear, and a snake, are crowded into bed with a little boy in the opening dream scene. As the familiar rhyme proceeds, the animals fall from the bed one by one. Each scene shows all ten inhabitants; some have found a new spot to sleep, and the remaining are still in or near the bed. On each double page, the "and one fell out" is clearly shown with one fewer character in the bed. Varying shades of blue and orange are effectively used to illustrate the dreamlike atmosphere.

Raffi. *Five Little Ducks.* Illustrated by Jose Aruego and Ariane Dewey. New York: Crown Publishers, 1989.

ISBN 0-517-56945-0	$12.00

★★ Single concept, PS–1

This beautifully illustrated picture book is a Raffi song adapted to print. Brightly colored drawings show a mother duck who seeks her ducklings as, one by one, they disappear. The "one less than" concept is a beginning look at the take-away model for subtraction. The young reader can use rational counting to verify the results. A happy ending occurs after Mother Duck searches for her young through the changing seasons. All five ducklings return as adults; the pictures show them with their mates and families, which range in size from one to five little ducks. Using the Raffi song from which the book was derived could add to the book's potential.

Rees, Mary. *Ten in a Bed.* Boston: Little, Brown & Co., 1988.

ISBN 0-316-73708-9 $13.95

★★ Single concept, PS–1

What a slumber party! Charming pastels illustrate this unique and humorous adaptation of the familiar counting rhyme "Ten in a Bed." The opening page shows ten little children crowded in a bed, and then the traditional rhyme begins. One by one the children fall out of bed. The details of the illustrations on the opposing pages show one more child joining the other wakened children who are involved in different activities around the house. These pages allow the reader to count the children who have been forced from the bed. This book introduces beginning addition and subtraction situations because the number in bed is constantly being decreased by one and the children humorously depicted on the opposite pages are always being joined by one more.

Richardson, John. *Ten Bears in a Bed: A Pop-up Counting Book.* New York: Hyperion Books for Children, 1992.

ISBN 1-56282-157-1 $13.95

★★★ Single concept, PS–K

This traditional rhyme and song has been interpreted many times in picture-book and storybook form. The pop-up format of this version affords a unique opportunity to see each character actually fall out of the bed! While counting down from 10 to 1 as the bears fall out of bed, young readers will also enjoy the antics of the other one to nine bears who are engaged in wonderful play activities. The illustrations are packed with details of their activities throughout the house, since other rooms can often be seen from an interesting overhead vantage point. The pop-ups are cleverly done; one example includes a bear actually rolling off the bed as a tab is pulled. Others are not as carefully designed; the tab is hard to pull on the page with three bears. The pop-ups clearly convey concepts involving beginning subtraction and counting back.

Samton, Sheila White. *The World from My Window: A Counting Poem.* Honesdale, Pa.: Boyd Mills Press, 1985.

ISBN 0-878093-15-0 $14.95

☆ Single concept, PS–1

"The moon is rising in the sky, and two pale clouds are drifting by." The opening lines of this counting poem set the stage with dreamlike illustrations and rhyming text that explore the number concepts from one through ten as they are represented in "the world from my window." Bold colors and interesting silhouettes combine to form attractive murallike illustrations. The numeral and word name for each set are simply stated under each illustration.

Thaler, Mike. *Seven Little Hippos.* Illustrated by Jerry Smath. New York: Simon & Schuster Books for Young Readers, 1991.

ISBN 0-671-72964-0 $13.95

☆ Single concept, PS–1

"Seven little hippos jumping on the bed. One bounced off and bumped her head." One by one, each of the seven little hippos falls after repeated escapades of jumping on the bed. Each time the doctor is called, and he offers the same advice: "No more little hippos jumping on the bed." The pastel watercolors are delightfully detailed and expressive. Whether hippos are crashing through the bedroom ceiling or falling into the goldfish bowl, each accident is novel and humorous. The story line is most appropriate for learning to count back or to subtract one. Children should enjoy chanting along with this appealing book, and the surprising ending is sure to get a giggle.

Wadsworth, Olive A. *Over in the Meadow: A Counting-Out Rhyme.* Illustrated by Mary Maki Rae. New York: Puffin, 1986.

ISBN 0-14-050606-3 $3.99
ISBN 0-590-44848-X   (Scholastic)   4.95

★★ Single concept, PS–K

Rich splashes of color in the earth tones of greens, blues, browns, and golds illustrate this traditional counting rhyme. A different animal mother is featured with an appropriately sized brood of little ones to illustrate the number concept on each double-page spread. Young readers will enjoy anticipating the actions and the rhyme of this charming story. The number word names alone are used in the text of the story, but the numerals are used in a haphazard way on the endpapers for decoration only.

West, Colin. *One Little Elephant.* Chicago: Candlewick Press, 1994.

ISBN 1-56402-375-3 $3.95

☆ Single concept, PS–2

Multicolored elephants, one after another, are added in amusing situations until there are ten altogether. The rhyming text anticipates each new elephant; for example, "Eight little elephants / Skipping in a line, / Once there were eight elephants, / And then there were...." The appropriate numeral is printed on each page as each new elephant appears in the clear, colorful drawings. West shows his creativity with a humorous ending and a second chance to count to ten.

_____. *Ten Little Crocodiles*. Chicago: Candlewick Press, 1995.

ISBN 1-56402-463-6 $3.95

☆ Single concept, PS–2

Ten whimsically drawn crocodiles disappear one at a time in a variety of ways. The last, lonely crocodile is surprised at the conclusion to find that the other nine are safe and sound as they are all reunited. A final numeral countdown from 10 to 1 is given on the last double page. Comical, colorful, and bright illustrations accompany a rhyming text. This is a companion book to *One Little Elephant*. West's humorous rhyming text and colorful, action-packed illustrations may become familiar favorites for young readers.

Number in Other Cultures

A major portion of this category is the Count Your Way series. Each book uses number as a vehicle for exploring another country and its culture. This series offers opportunities for integrating mathematics and social studies. Other books in this category, Feelings's *Moja Means One,* Trinca and Argent's *One Woolly Wombat,* and Hartmann's *One Sun Rises: An African Wildlife Counting Book,* provide a more useful balance between the two subjects.

Claudia Zaslavsky's *Count on Your Fingers African Style* is out of print but is considered exceptional. A copy may be available in your library. See the first edition of *The Wonderful World of Mathematics* for an annotation.

Brett, Jan. *The Twelve Days of Christmas.*
See Early Number Concepts: Chants, Songs, and Rhymes

Dee, Ruby. *Two Ways to Count to Ten.* Illustrated by Susan Meddaugh. New York: Henry Holt, 1988.

ISBN 0-8050-0407-6	$14.95
ISBN 0-8050-1314-8 (paper)	5.95

★★ Single concept, K–3

In this retelling of a traditional Liberian folktale, King Leopard decides to hold a contest to find his successor. He challenges all the beasts in the jungle to throw a spear so high that they can count to ten before it lands. One after another the animals try and fail, until a clever young antelope uses a different way to count to ten—counting by twos. The expressive illustrations complement the mood and the actions of the characters as the story unfolds. Rote counting up to nine is repeated throughout the book. The real mathematical strength lies in using this story as an introduction to exploring other ways to count. The book's main strength is the introduction of folklore from other cultures.

Feelings, Muriel. *Moja Means One: A Swahili Counting Book.* Illustrated by Tom Feelings. New York: Dial Books for Young Readers, 1976.

ISBN 0-8037-5711-5   (paper)   4.95
ISBN 0-14-055296-0   (Puffin Books, 1994)   18.99
ISBN 0-14-054662-6   (paper, Puffin Books, 1992)   4.99

★★★ Single concept, PS–3

This beautifully illustrated book successfully reinforces counting skills by furnishing opportunities to count objects, one through ten. The memorable illustrations introduce the reader to the uniqueness of an East African people and the area in which they live. Swahili names for the numbers through ten are presented in the text. The serenity of the landscape and the dignity of the people are emphasized in the soft gray charcoal drawings.

Grossman, Virginia, and **Sylvia Long.** *Ten Little Rabbits.* San Francisco: Chronicle Books, 1991.

ISBN 0-87701-552-X $12.95
ISBN 0-8118-1057-7   (paper)   6.95

★★ Single concept, PS–2

Beautiful pastel illustrations depict the environment, lifestyles, and traditions of ten Native American cultures through the activities of rabbits dressed in native costumes. On the desert soil with an adobe home in the background, two rabbits, dressed as graceful Hopi dancers, ask for rain. On a barren, snow-covered plain under a starry sky, five rabbits, wrapped in Indian blankets and huddled around a fire, are depicted as five wise storytellers. On a final double page, each culture—Sioux, Tewa, Ute, Menominee, Blackfoot, Hopi, Arapaho, Nez Percé, Kwakiutl, and Navajo—is introduced and linked to the appropriate illustration as the numerals from 1 through 10 appear. An illustration of a typical design for an Indian rug appears beside a brief description of each culture. This is a well-designed counting book whose simple text and charming illustrations combine to provide a fine balance between the introduction of Native American cultures and the opportunity to explore counting. Children should find this book appealing.

Hartmann, Wendy. *One Sun Rises: An African Wildlife Counting Book.* Illustrated by Nicolaas Maritz. New York: Dutton Children's Books, 1994.

ISBN 0-525-45225-7 $13.99

★★★ Single concept, PS–1

Bold illustrations reflect the power of one brilliant African sun as it rises in the clear, blue sky. Vibrant colors and black outlines define the native animals—two kestrels, three elephants, four beetles, five suricates—depicted in their day environment in groups of one through ten. A transition takes place as "Ten vultures wait" as the sun goes down. A brief moment of calm is captured on the next double-page spread as the moon casts its silver light on the land beneath. The arrival of the animals of the night proceeds as "Ten reed frogs praise the moon," "Nine moths flutter in her light," ..., and finally, "One sun rises over Africa." This circle story, with its rhythmic text and plot, is strangely poetic as it captures the calm, the beauty, and the sometimes ominous presence of the hunter and the hunted. The opportunity to count objects up through ten and then down from ten is merely the vehicle through which a perfectly matched author and illustrator have woven their magic. A paragraph with more information about each featured animal is included at the back of the book.

Haskins, Jim. *Count Your Way through Africa.* Illustrated by Barbara Knutson. Minneapolis, Minn.: Carolrhoda Books, 1989.

ISBN 0-87614-347-8 $18.95
ISBN 0-87614-514-4   (paper)   5.95

☆ Single concept, 1–4

All the books in the Count Your Way through... series have a common format. Colorful illustrations enhance the texts, which integrate history, geography, and number. Each number from one through ten is introduced with the respective country's corresponding symbol and its pronunciation. Each number is related to information about the culture, geography, and history of that country, which could motivate the reader to pursue further research on the country. A history of the country's language and alphabet is included in the introductory notes. The main purpose of these books is to introduce the reader to other cultures and they do so only superficially. This series should not be viewed as books to develop counting or other number concepts; number is simply a vehicle for an introduction to various nations in the world. Often the number of objects that could be counted is not directly related to the text, or there are no objects to count.

In *Count Your Way through Africa,* Swahili, one of 800 languages in Africa, is the language used to count from one through ten. The informative text is softly complemented by the brown-toned watercolor illustrations.

———. *Count Your Way through Canada.* Illustrated by Steve Michaels. Minneapolis, Minn.: Carolrhoda Books, 1989.

ISBN 0-87614-350-8 $18.95
ISBN 0-87614-515-2   (paper)   5.95

☆ Single concept, 1–4

French, the official language of the Canadian province of Quebec, is highlighted in writing the number words from one through ten.

The number two is used to introduce the fact that Canada has two official languages. As in the other books in this series, number concepts are interwoven into the otherwise informative text in a contrived fashion; for example, a traditional Inuit winter costume has eight pieces of clothing—that is, if each mitten and boot is counted separately. Canadian geography, history, and culture facts are featured throughout the text and complemented with satisfactory artwork. See the annotation for *Count Your Way through Africa* for an overview of this series.

————. *Count Your Way through China.* Illustrated by Dennis Hockerman. Minneapolis, Minn.: Carolrhoda Books, 1987.

ISBN 0-87614-302-8   (library binding) $18.95
ISBN 0-87614-486-5   (paper)   5.95

☆ Single concept, 1–6

Count Your Way through China explores the fact that written Chinese has no alphabet but uses individual characters to stand for syllables or whole words. The author shows how to write and pronounce the numbers from one through ten in Chinese. Each number is then used as a vehicle to explore Chinese history, geography, and lifestyle. Often the examples for the numbers are forced; for the example "Three Chinese *li* equal a mile," the illustration shows three people walking along the Great Wall. Rich watercolor illustrations enhance the text. See the annotation for *Count Your Way through Africa* for an overview of this series.

———— *Count Your Way through India.* Illustrated by Liz Brenner Dodson. Minneapolis, Minn.: Carolrhoda Books, 1990.

ISBN 0-87614-414-8 $18.95
ISBN 0-87614-577-2   (paper)   5.95

☆ Single concept, 1–4

In *Count Your Way through India,* introductory notes indicate that Hindi, an official language of India, and English are used for most governmental business in this land of 800 million people and fifteen major languages. The Hindi alphabet is used within the book to name the numbers from 1 through 10; the pronunciation guide offers a challenge with two "r"s to represent the unique rolling "r" of the Hindi language. Numbers are the vehicle to introduce India, from one banyan tree, which can look like a whole forest; two circles on the flag; three items of furniture kept by Gandhi; four tall towers or minarets on the Taj Mahal; ..., to the ten-headed figure of Ravana. Delicately designed top and bottom borders effectively frame the dreamy illustrations and the corresponding text. Dodson's illustrations convey the mystery, beauty, and legends of the Indian culture. See the annotation for *Count Your Way through Africa* for an overview of this series.

————. *Count Your Way through Israel.* Illustrated by Rick Hanson. Minneapolis, Minn.: Carolrhoda Books, 1990.
ISBN 0-87614-415-6 $18.95
ISBN 0-87614-558-6   (paper)   5.95
☆ Single concept, 1–4

Background information on the establishment of the state of Israel as the Jewish homeland is found in the introductory notes, which also include a discussion of the Hebrew language of twenty-six characters, all consonants. The first ten letters of the Hebrew alphabet are also used for the numerals 1 through 10; the corresponding numeral-letter is also integrated into the borders that frame each page. Number is the vehicle to introduce elements of this culture: a map illustrates the one country that is Israel; the biblical story of David and Goliath portrays two men; the holiday of Passover includes four questions asked at the seder meal; an official symbol is the menorah with its seven branches; a wildlife program protects eight endangered species. Hanson's airbrush and watercolor illustrations enhance this introductory portrayal of the diversity that is Israel. See the annotation for *Count Your Way through Africa* for an overview of this series.

————. *Count Your Way through Italy.* Illustrated by Beth Wright. Minneapolis, Minn.: Carolrhoda Books, 1990.
ISBN 0-87614-406-7 $18.95
ISBN 0-87614-533-0   (paper)   5.95
☆ Single concept, 1–4

Introductory notes for this volume of the series are brief and comment on only the Italian language and its pronunciation. Numbers from one through ten are again the thread for tales intertwining Italy's legends, culture, and geography: a giant named Typhon was buried under one volcano, Mount Etna; Rome was named after one of two brothers, Romulus and Remus; Columbus sailed in three ships; theater in the sixteenth to the eighteenth centuries was performed with a cast of eight characters. Wright's bold vibrant illustrations add to the sampling of Italian culture of this counting tour of Italy. See the annotation for *Count Your Way through Africa* for an overview of this series.

————. *Count Your Way through Japan.* Illustrated by Martin Skoro. Minneapolis, Minn.: Carolrhoda Books, 1987.
ISBN 0-87614-301-X   (library binding) $18.95
ISBN 0-87614-485-7   (paper)   5.95
☆ Single concept, 1–6

After an introductory note about the history of Japanese numbers and how some numbers are Chinese derivations, the book relates each number to information about the culture, geography, and history of Japan. Counting tasks, as in other books in this series, are often contrived and are of secondary value. The format of using numbers to tell about a country is unique and interesting, but the value of the book lies in the exploration of another culture, not in the mathematics. Inconsistent examples of number include a discussion on the new responsibilities and privileges that children acquire when they reach the age of ten, but the picture shows ten children. Another example tells of the six major sumo tournaments but depicts six wrestlers. The delicate watercolor illustrations support the author's intent to initiate a better understanding of Japan, its people, and its culture. See the annotation for *Count Your Way through Africa* for an overview of this series.

————. *Count Your Way through Korea*. Illustrated by Dennis Hockerman. Minneapolis, Minn.: Carolrhoda Books, 1989.

| ISBN 0-87614-348-6 | | $18.95 |
| ISBN 0-87614-516-0 | (paper) | 5.95 |

☆ Single concept, 1–4

Sino-Korean numbers are used to count minutes and money, but Korean numbers, which are used to count people and things, are the ones introduced in this book. Attractive, vibrant illustrations and an informative text highlight facts about Korea, which include one famous astronomical observatory, a pagoda five stories high, and ten vowels in the Korean alphabet. See the annotation for *Count Your Way through Africa* for an overview of this series.

————. *Count Your Way through Mexico*. Illustrated by Helen Byers. Minneapolis, Minn.: Carolrhoda Books, 1989.

| ISBN 0-87614-349-4 | | $18.95 |
| ISBN 0-87614-517-9 | (paper) | 5.95 |

☆ Single concept, 1–4

Mexico, one country in North America, is featured in this book from the *Count Your Way through...* series. The number words from one through ten are written in Spanish, the official language of Mexico. Traditions and history from Indian and Spanish cultures are interestingly woven through the text. Beautiful pastel illustrations clearly depict the ideas presented. See the annotation for *Count Your Way through Africa* for an overview of this series.

————. *Count Your Way through Russia*. Illustrated by Vera Mednikov. Minneapolis, Minn.: Carolrhoda Books, 1987.

| ISBN 0-87614-303-6 | (library binding) | $18.95 |
| ISBN 0-87614-488-1 | (paper) | 5.95 |

☆ Single concept, 1–6

Each number from one through ten is used to relate information about Russian history, geography, or traditions. The reader is shown how to pronounce the numbers written in the Cyrillic alphabet of the Russian language and how to write the numbers with Arabic numerals. Some examples are contrived, such as the four days needed to go halfway across the Soviet Union on the Trans-Siberian train to represent 4; others, such as the seven nesting dolls called Matryoshka to represent 7, are most appropriate. This book could leave the reader wanting to do more research on the country. See the annotation for *Count Your Way through Africa* for an overview of this series.

————. *Count Your Way through the Arab World.* Illustrated by Dana Gustafson. Minneapolis, Minn.: Carolrhoda Books, 1987.

ISBN 0-87614-304-4	(library binding)	$18.95
ISBN 0-87614-487-3	(paper)	5.95

☆ Single concept, 1–6

Travel through the Arab world and learn about its culture, religion, and geography. Numbers from one through ten are used in the informative text and complementary illustrations. The reader can learn how to pronounce the numbers in the Egyptian dialect and how to write the numbers by using the Arabic symbols. The book integrates number with the history and geography of the Arab world. See the annotation for *Count Your Way through Africa* for an overview of this series.

Haskins, Jim, and **Kathleen Benson.** *Count Your Way through Brazil.* Illustrated by Liz Brenner Dodson. Minneapolis, Minn.: Carolrhoda Books, 1996.

ISBN 0-87614-873-9		$18.95
ISBN 0-87614-971-9	(paper)	5.95

☆ Single concept, 1–4

An introductory note about Brazil warns of the difficulty of making generalizations about a country so large and diverse. Although the official language is Portuguese, the infusion of native Indian and African languages has created new forms of pronunciations, which makes the versions of Portuguese spoken here unique to Brazil. This cultural counting story introduces Brazil as one nation with five regions, seven different ethnic groups, nine important products from trees, and ten instruments to play the music of the country. Haskins and Benson are coauthors of this book, but see the annotation for Haskin's *Count Your Way through Africa* for an overview of this series.

————. *Count Your Way through France.* Illustrated by Andrea Shine. Minneapolis, Minn.: Carolrhoda Books, 1996.

ISBN 0-87614-874-7 $18.95
ISBN 0-87614-972-7   (paper) $5.95

☆ Single concept, 1–4

Introductory notes provide information on the French language and the influence of English words on popular language. Again, numbers from 1 through 10 are used as a vehicle for a brief glimpse into various aspects of French life: one familiar symbol, the Eiffel tower; two international races, the Tour de France and Le Mans; three stripes of the flag, the tricolors; nine foods, including famous French cuisine like mousse and croissants. Shine's charming watercolors do much to bring the informative text alive. Haskins and Benson are coauthors of this book, but see the annotation for Haskin's *Count Your Way through Africa* for an overview of this series.

————. *Count Your Way through Greece.* Illustrated by Janice Lee Porter. Minneapolis, Minn.: Carolrhoda Books, 1996.

ISBN 0-87614-875-5 $18.95
ISBN 0-87614-973-5   (paper)   5.95

☆ Single concept, 1–4

Introductory notes provide limited information except to acknowledge the power and influence of Greek culture on the Western world: language, system of government, literature, architecture, and sculpture. From one official religion to ten animals featured in Aesop's fables, the informational text unfolds accompanied by Porter's appealing illustrations. Haskins and Benson are coauthors of this book, but see the annotation for Haskin's *Count Your Way through Africa* for an overview of this series.

————. *Count Your Way through Ireland.* Illustrated by Beth Wright. Minneapolis, Minn.: Carolrhoda Books, 1996.

ISBN 0-87614-872-0 $18.95
ISBN 0-87614-974-3   (paper)   5.95

☆ Single concept, 1–4

Introductory notes tell of a divided Ireland, which in spite of the differences between the north and the south, is bound by a common history, language, and culture. Number words are in Irish Gaelic; an interesting fact is that although Gaelic is one of the official languages, it may be pronounced differently in various parts of Ireland. No indication is given about which dialect is used for the pronunciation of the number names. From one Saint Patrick to five lines in a limerick to ten handcrafted goods, the reader is introduced to ten aspects of Ireland. Wright's illustrations are warm and dreamlike and appeal to the eye. Haskins and Benson are coauthors of this book, but see the annotation for Haskin's *Count Your Way through Africa* for an overview of this series.

Linden, Ann Marie. *One Smiling Grandma: A Caribbean Counting Book.* Illustrated by Lynne Russell. New York: Dial Books for Young Readers, 1992.

ISBN 0-8037-1132-8 $15.00

★★ Single concept, PS–K

On the basis of her childhood memories of Barbados, Linden introduces the numbers from one through ten through a young girl's observations of her Caribbean island. Each situation is described in rhyming text: "One smiling grandma in a rocking chair, / Two yellow bows tied in braided hair." Russell's lush watercolors capture the beauty of the tropical island and its people. Illustrations containing four steel drums, five flying fish, six market ladies, and seven conch shells are easy to count; numerals are not included. Through both verse and illustrations the reader is given opportunities to count, as well as stunning images of the island.

Mora, Pat. *Uno, dos, tres: One, Two, Three.* Illustrated by Barbara Lavallee. New York: Clarion Books, 1996.

ISBN 0-395-67294-5 $14.95

★★ Single concept, K–2

Mexican markets provide the theme for this counting book, which includes the Spanish names for the numbers from one through ten. Two young girls search the market to buy Mexican artifacts or other items, such as *uno,* or one, Aztec sun; *dos,* or two, doves; *cuatro,* or four, pinatas; or the *ocho,* or eight, woven baskets. More than one example is given for different numbers, and opportunity to practice all the word names in order is provided. The attractive mural-like illustrations depict the activity and color of a lively Mexican marketplace.

Onyefulu, Ifeoma. *Emeka's Gift: An African Counting Story.* New York: Cobblehill Books, 1995.

ISBN 0-525-65205-1 $14.99

★★★ Single concept, PS–2

Emeka's photograph on the front cover is a smiling welcome and an invitation to walk with him to his grandmother's house and explore some important aspects of his Igalan culture. The narrative describing Emeka's walk and his thoughts on what to buy for his grandmother is contained in a brightly bordered box on each page; informative explanations and expansions on what is seen in the brilliant photographs are included in a box with a plain outline. The author's introductory note sets the stage for this pleasant counting story of one boy, who meets two friends playing *okoso,* greets three women on their way to the market, admires four brooms on display at the market—all part of a photographic journey to his beloved grandmother's house. The cultural experience is far more important than the counting possibilities. Numerals and number words are set in bold text; the sets are easy to count and each presents another dimension of the culture of the Igala people from southern Nigeria. This book is a good choice for introducing the reader to other cultures or for integrating theme work in the classroom.

Trinca, Rod, and **Kerry Argent.** *One Woolly Wombat.* Illustrated by Kerry Argent. Brooklyn, N.Y.: Kane/Miller Book Publishers, 1987.

ISBN 0-916291-00-6 $12.95
ISBN 0-916291-10-3   (paper)   6.95

★★★ Single concept, K–2

One woolly wombat, two cuddly koalas, three warbling magpies, and four thumping kangaroos are some of the Australian animals populating this delightful counting book, which includes numbers from one through fourteen. A phrase such as "five pesky platypuses splashing with their feet" introduces each double-page spread. The scene that includes a platypus in an inner tube is humorously depicted on the opposite page in beautiful illustrations. The next double page continues in rhyming text such as "six cheeky possums looking for a treat." By clever design throughout the book, the scenes appear three-dimensional with the animals ready to burst forth from the page.

Counting—Zero

Zero is a unique number that needs a special introduction. Children develop the concept of no objects in a set after they develop the concept of how many in a set, using only the natural numbers. The counting books in this category introduce the concept of zero: the numeral, the word name, and empty sets.

Anno, Mitsumasa. *Anno's Counting Book.* New York: HarperCollins Children's Books, 1977.
ISBN 0-690-01287-X $16.00
ISBN 0-690-01288-8   15.89
★★★ Multiconcept, PS–3

Beautiful watercolor landscapes show the changing seasons through the twelve months of the year and the activities of the people and animals who live there. The wordless picture book depicts the numbers zero through twelve. Each double page provides a number of sets for the reader to find and to count, such as six buildings, six geese, six children, six adults, six evergreens, and six deciduous trees, and introduces the numeral. As the items in the illustrations change in number, a growing tower of cubes illustrates zero through ten. A second tower of cubes is used to illustrate eleven and twelve as ten and one and ten and two, respectively. This illustration provides a simple introduction to place-value concepts. A section at the conclusion of the book is meant to assist teachers or parents in explaining the concepts illustrated in the text.

Aylesworth, Jim. *One Crow: A Counting Rhyme.* Illustrated by Ruth Young. New York: HarperCollins Children's Books, 1990.
ISBN 0-0-06-443242-4 $5.95
★★ Multiconcept, PS–1

The sun is rising as another busy summer day on the farm begins. This beautifully illustrated counting book features barnyard animals in sets from zero through ten with corresponding numerals. Each easy-to-count set of animals is introduced through rhyming verse. From sunrise to sunset, the day is filled with sets of animals actively involved with their daily routines in their natural surroundings. Then summer turns to winter, and the barnyard is transformed into a cold and snowy world by a blanket of white snow. The numbers from zero through ten are revisited, and the rhymes reflect the seasonal change experienced by each set of barnyard animals in their winter wonderland. This is a good book for reading and discussing with children as they compare the changes between the two sections of the book.

Bowen, Betsy. *Gathering: A Northwoods Counting Book.* Boston: Little, Brown & Co., 1995.

ISBN 0-316-10371-3 $15.95

☆ Single concept, PS–3

Preparation for winter in the Northwoods is the theme for introducing the numbers from zero through twelve. Spring includes such garden activities as producing each vegetable from just one seed or fixing two rhubarb pies in June. Summer includes forming three memories with friends, preparing for hibernation by a mother bear and her three cubs, harvesting blueberries and wild rice, or fishing for walleyes. Objects to count, the numeral, and the word name appear with each description. The numbers are sequentially introduced; the examples selected for different numbers are arbitrary, for example, five blueberries, six bags of rice, and nine extension cords. "Zero degrees" is the title of a page depicting zero; rather than depict the absence of any objects, connections are made to winter experiences. A spring scene is illustrated; the text discusses that winter seems a long way off and refers to the changes that will take place as the temperature plunges to zero. This example would not help the reader develop the concept of zero. Another example shows eight logs to count, but the caption is "Eight *cords* of wood." Detailed descriptions of each Northwoods activity are complemented by wood-block prints. The main purpose of this book is to introduce the reader to the environment of the Northwoods.

Gould, Ellen. *The Blue Number Counting Book.* Illustrated by Cathy Kelly. Washington, D.C.: Learning Tools, 1983.

ISBN 0-938017-01-2   (paper) $6.00

★★ Single concept, PS–2

Numbers from zero through ten are presented as felt numerals and felt cutouts of objects to appeal to the child's sense of touch. The only color used in the predominantly black-and-white drawings is blue, and both the numerals and the appropriate number of objects are shown in blue felt. This is a unique opportunity for the child to feel the numeral and the object while counting and saying the name for the number.

Jackson, Woody. *Counting Cows.* San Diego, Calif.: Harcourt Brace & Co., 1995.

ISBN 0-15-220165-3 $14.00

★★ Single concept, PS–3

Black-and-white bovines abound in this unusual counting book, which sequentially features the numbers from ten through zero. Both numerals and number word names are included; the minimal, but lyrical, text includes descriptions like "nine fenced Friesians," "six haying heifers," "two crossing bossies," and "one big bovine." In the scene described as "seven sunning cows," seven black-and-white cows are shown in a landscape painted with brilliant tones of blue, orange, magenta, gold, and green. The media for the attractive illustrations are gouache and watercolor. After seeing "zero" cows in the starry rural landscape, the reader will encounter the surprise ending—all ten cows dressed for, and participating in, a barn dance!

Kosowsky, Cindy. *Wordless Counting Book.* Bridgeport, Conn.: Greene Bark Press, 1992.

ISBN 0-880851-00-8 $10.95

★★★ Single concept, PS–K

Eleven bright colors divide the book's front and back covers into eleven vertical panels, which contain the numerals from 0 through 10. As the reader opens the book, the same view of the numerals appears with the corresponding rainbow colors. Each sequential cardboard page increases in length, which creates a layered effect. The shortest turquoise page shows the numeral 0. As the reader turns this narrow page, he or she sees the entire green panel with an illustration of one green apple next to the numeral 1. The next wider page has an illustration of two oranges next to the numeral 2. Across the bottom of this orange page is a small illustration of the one green apple with the numeral 1. This pattern builds so that on the last page, which shows ten slices of watermelons next to the numeral 10, the bottom panel pictures 1 green apple, 2 oranges, 3 yellow lemons, 4 blueberries, 5 red strawberries, 6 purple grapes, 7 gold brown pineapples, 8 yellow bananas, and 9 cherries with the corresponding numerals. Number word names are not included. The book's clever design displaying the increasing size of the numbers, the inclusion of 0, and the repeating border sequence are very well done. Additionally, the bright colors, the simple illustrations that are easy to count, and the book's sturdy construction provide a winning combination.

Pallotta, Jerry. *The Icky Bug Counting Book.* Illustrated by Ralph Masiello. Watertown, Mass.: Charlesbridge Publishing, 1992.

See Early Number Concepts: Counting—Large Numbers

———. *Cuenta los insectos.* Illustrated by Ralph Masiello. Watertown, Mass.: Charlesbridge Publishing, 1993.

See Early Number Concepts: Counting—Large Numbers

Salt, Jane. *My Giant Word and Number Book.* Illustrated by Sarah Pooley. New York: Kingfisher Books, 1992.

See Early Number Concepts: Counting—Other Number Concepts

Winik, J. T. *Fun with Numbers.* Milwaukee: Durkin Hayes, 1985.

ISBN 0-88625-104-4   (paper) $2.95

★★ Single concept, PS–1

Children of different races are involved in playful activities, with each new activity attracting another child. The book offers its readers the opportunity to count the children and to identify the representative numerals. The appropriate large numeral and the written number word are included on each page. At the end of the book when all the children leave for lunch, the empty tire swing represents a big zero. In one example, ordinal concepts are presented and written as numerals. The illustrations are bold and inviting, with no distractions.

Counting—
Large Numbers

Counting numbers larger than ten soon becomes a challenge for kindergarten children. They love to practice a newly learned skill, and many of these books will give the opportunity to do just that. As in the simpler counting books, the variety of forms in which the challenge is offered is limitless. Whether embedded in a fascinating story, contained within a listing of items, or captured in the magic of a catchy rhyme, the challenge is there.

Some of these books introduce each consecutive counting number, like Tucker's *1 2 3 Count with Me,* which represents numbers from one through twenty, or Grover's *Amazing and Incredible Counting Stories* with numbers from one through twenty five. Some, such as Reiss's *Numbers,* represent decade numbers after introducing the numbers from one through twenty. When decade numbers are introduced, the sets may not be arranged in groups of ten. Questions that motivate children to consider benchmarks and to estimate will help these readers develop number sense. Books introducing benchmarks include Thornhill's *The Wildlife 1, 2, 3: A Nature Counting Book,* featuring 25, 50, 100, and 1000, and Wood's *One Tortoise, Ten Wallabies: A Wildlife Counting Book,* featuring 20, 25, 50, and 101! Illustrations depicting the relative size of numbers, like those in Conran's *My First 1, 2, 3 Book* should also enhance the reader's number sense.

The following books are out of print but are considered exceptional:

Brown, Marc. *One Two Three: An Animal Counting Book*

Demi. *Demi's Count the Animals 1-2-3*

Copies may be available in your library. See the first edition of *The Wonderful World of Mathematics* for annotations.

Cave, Kathryn. *Out for the Count: A Counting Adventure.* Illustrated by Chris Riddell. New York: Simon & Schuster Books for Young Readers, 1992.

See Early Number Concepts: Counting—Place Value

Cole, Norma. *Blast Off! A Space Counting Book.* Illustrated by Marshall Peck III. Watertown, Mass.: Charlesbridge Publishing, 1994.

ISBN 0-88106-498-X   (softcover) $6.95
ISBN 0-88106-499-8   (trade)   14.95
ISBN 0-88106-493-9   (library binding)   15.88

☆ Single concept, PS–4

In the opening pages, the traditional rhyme "One two, buckle my shoe" has been adapted to accommodate a space theme. Like the traditional rhyme, this narrative is merely a means for rote counting; the number does not relate to the space scene. Placed at the top of the page, the rhyme contains only the number word reference and a suggested action. A numeral from 1 through 20, accompanied by an appropriate number of space objects, appears on each page. These objects are clustered next to the numeral in a variety of ways, including arrays, triangular numbers, and groups of ten. Zero is presented first, before the space adventure begins, and the presentation is confusing because of a lack of context. A double-page spread introduces the numerals and word names for the decade numbers to one hundred and the increasing larger place values from one thousand to one trillion. The remaining section of the book depicts counting backward from ten to zero. At the foot of each page, an informative text in small print supplies interesting space-related information about the magnificent illustrations. Brilliant color reflects the intricacies of the space ship and the expressions of the three young astronauts, but more important, creates a feeling of awe at the beauty and activity of the expansive firmament.

Conran, Sebastion. *My First 1, 2, 3 Book.* New York: Simon & Schuster Children's Books, 1988.

ISBN 0-689-71267-7 $7.95

★★ Multiconcept, PS–1

Brightly colored double-page spreads illustrate the numbers from 1 through 20. A minimal text complements the design: "three cats ... prowling," or "ten soldiers ... at attention." A panel on the right-hand page highlights the numeral and an appropriate number of marble counters. The counters are arranged in arrays or partial arrays for ease in counting. Since the counters for 4 through 10 are arranged in columns of 2, the arrangement clearly depicts odd and even numbers. At 20 the counting pattern changes, and the numbers 30, 40, 50, 60, 100, 1000, and 1 000 000 are depicted. To accommodate the larger numbers, the counters in the border are smaller for the multiples of 10 and increasingly smaller for 100, 1000, and

1 000 000. The final scenes are obviously uncountable but simply illustrate the largeness of the numbers.

Eichenberg, Fritz. *Dancing in the Moon: Counting Rhymes.* New York: Voyager Books, 1975.

ISBN 0-15-623811-X   (paper) $3.95

★★ Single concept, K–3

Originally published in 1955, this delightful counting book continues to appeal to the reader. The numbers from one through twenty are introduced in rhymes: "4 pandas resting on verandas, 5 dragons pulling wagons." Eichenberg's clever illustrations and choice of rhyming words add humor to the counting. The detailed pastel drawings should enhance the reader's interest.

Fleming, Denise. *Count!* New York: Henry Holt & Co., 1992.

See Early Number Concepts: Counting—Place Value

Grande Tabor, Nancy Maria. *Cincuenta en la Cebra: Contando con los Animales. / Fifty on the Zebra: Counting with the Animals.* Watertown, Mass.: Charlesbridge Publishing, 1994.

See Early Number Concepts: Counting—Other Number Concepts.

Grover, Max. *Amazing and Incredible Counting Stories.* San Diego, Calif.: Browndeer Press, Harcourt Brace & Co., 1995.

ISBN 0-15-200090-9 $14.00

☆ Single concept, K–3 (2–4)

"Skyscrapers Missing, 1 Found," "11 Telephones Found Growing in Woods," and "14-Piece Spoon Necklace Shocks Fashion World" are some of the zany headlines introducing the numbers from 1 through 25, 50, 75, and 100. An equally incredible news story and illustration accompany each headline: one of the missing skyscrapers is found near the edge of town by a search party; as hikers arrive on the scene, the 11 telephones hanging from the tree stop ringing, since the caller had just hung up; a fashion model pauses on the runway so the onlookers, who are eating their soup with forks, can admire her necklace made with 14 spoons. Some of the situations introduce two numbers at once, such as "22 Scientists Stumped by 23-Piece Puzzle." As in other books that introduce large numbers, some of the sets are randomly arranged, which makes them more difficult to count. But the page illustrating 75 clocks is relatively easy to count because of the placement of the clocks on the castle walls. Numerals rather than number word names are included. Intense color illustrations, executed in a flat, cartoonlike style, express the wild humor and the unlikely actions conveyed in the text. Grover's wild tales are sure to entertain, although the zany "headline" humor may escape the young reader. A most appropriate final headline encourages both reading and counting by proclaiming, "Millions of Readers Learn by Counting Stories."

Johnson, Odette, and **Bruce Johnson.** *One Prickly Porcupine.* New York: Oxford University Press, 1991.
See Early Number Concepts: Counting—Partitions

Kitchen, Bert. *Animal Numbers.* New York: Dial Books for Young Readers, 1987.

ISBN 0-8037-0459-3		$12.95
ISBN 0-8037-0910-2	(paper)	4.95
ISBN 0-14-054602-2	(paper, Puffin)	5.95

☆ Single concept, PS+

"Answer me this / If you're in the mood, / How many babies / In each mother's brood?" Clear, warm watercolors of intricately defined animal portraits provide an elegant counting book for observant young readers. Similar in design to those in Kitchen's equally distinctive alphabet book, each symbol is represented in bold, black print with the animals intertwined. The numerals from 1 through 10 are presented, then 15, 25, 50, 75, and 100. An interesting feature of the painting for 100 is that the readers, should they choose to count so many, are asked to count frogs from the frog's eggs through the growing tadpoles to the frog herself. The final page includes interesting information on the animals featured.

The Lifesize Animal Counting Book. New York: Dorling Kindersley, 1994.
See Early Number Concepts: Counting—One to Ten

Lottridge, Celia Barker. *One Watermelon Seed.* Illustrated by Karen Patkau. New York: Oxford University Press, 1990.
See Early Number Concepts: Counting—Place Value

MacCarthy, Patricia. *Ocean Parade.* New York: Dial Books for Young Readers, 1990.

ISBN 0-8037-0780-0 $11.95

☆ Single concept, PS–2

Schools of fanciful fish swim across the pages to introduce the numbers from 1 through 20, then the numbers 30, 40, 50, and 100. More than one number concept is introduced on most pages; for example, "3 flat fish / 4 thin fish / 5 fat fish" are clearly depicted as each set of fish is distinctly illustrated through size, color, pattern, and shape. Larger numbers, such as the 17 red and white spotted fish and the 18 blue and green striped fish, are easily classified, but the counting will become more challenging for the young child. Such examples as the 40 golden fish or the 50 pink and green spotted fish could be used to encourage the development of number sense for large numbers. Experienced counters will find these a challenge, since no pattern or clustering is present to make counting easier. The author created the colorful illustrations by using batik paintings on silk.

Micklethwait, Lucy. *I Spy Two Eyes: Numbers in Art.* New York: Greenwillow Books, 1993.

ISBN 0-688-12640-5		$19.00
ISBN 0-688-12642-1	(library binding)	18.93

★★ Single concept, PS+ (all ages)

Micklethwait offers an understandable explanation for her choice of twenty magnificent works of art as a focus for a counting activity. Building on her experiences with her own children, she recognizes children's love of detail and challenges them with the familiar "I spy" game, which opens each page. "I spy one fly"; immediately the reader's eyes move to the magnificent reproduction of *Portrait of a Woman of the Hofer Family* to check to see if she or he can spy it, too. Subsequent pages spy two eyes, three puppies, four fish, ..., twenty angels in beautiful reproductions of masterpieces by such artists as Appel, Gauguin, Matisse, and Botticelli. The challenge of the "I spy" activity is not always easily met. Many items are difficult to find and demand great attention to detail. The challenge has great rewards because not only do readers find what they are looking for, they find much more, especially a growing appreciation of the artists themselves. Number words and numerals are highlighted by large print on plain white, heavy gloss paper with the simple opener for each page. At the bottom of this page, the name of the artist and the title of the work are indicated; additional credits on the closing pages indicate further information, such as the artist's time period and the current location of the original artwork.

My First Look at Numbers. New York: Random Books for Young Readers, 1990.

ISBN 0-679-80533-8 $8.00

★★ Single concept, PS–K

Crisp photographs make the appealing objects easy to count in this beginning counting book. Colorful objects are photographed against a white background, and bold numerals show the numbers from one through ten. Larger sets of 20 crayons and 100 candies are also included. The latter photograph is particularly striking because the different candies are arranged in rows of 10 in a simple pattern. Number word names are included in the simple text.

My First Number Book. New York: Dorling Kindersley, 1992.
See Number—Extensions and Connections: Number Concepts

Pallotta, Jerry. *The Icky Bug Counting Book.* Illustrated by Ralph Masiello. Watertown, Mass.: Charlesbridge Publishing, 1992.

ISBN 0-88106-497-1		$14.95
ISBN 0-88106-690-7	(library reinforced)	15.88
ISBN 0-88106-496-3	(paper)	11.95

★★ Single concept, PS–3

Number concepts from zero through twenty-six are introduced in *The Icky Bug Counting Book*. Large numerals and number word names with an appropriate number of bugs to count are included in each layout. A paragraph about each highlighted bug includes facts about its habitat, appearance, and special characteristics. Sometimes the information conveys the fact that some "bugs" are not bugs at all. Pillbugs, for example, are crustaceans. The text is easy to read and conversational in tone. Most illustrations show the bugs with their protective coloration against a background of their habitat, such as tree branches or rocks. In some examples, such as the eleven pillbugs, fourteen millipedes, and twenty-three longlegs, the camouflage is so well done that the objects become very difficult to count. A dog is pictured for the example of twenty-one fleas with an explanation that the dog has that many fleas, but the fleas are so tiny that they cannot be readily seen or counted. An added surprise at the book's end is an explanation of why the book ends at the number twenty-six. The letters in the alphabet are represented backward from one zebra swallowtail and two yellowjacket flies to twenty-five blister beetles and twenty-six army ants. The only exception is the third picture of three elegant crab spiders.

————. *Cuenta los insectos*. Illustrated by Ralph Masiello. Watertown, Mass.: Charlesbridge Publishing, 1993.

ISBN 0-88106-639-7	(library reinforced)	$15.88
ISBN 0-88106-419-X	(paper)	6.95

★★ Single concept, PS–3

Cuenta los Insectos is the Spanish version of *The Icky Bug Counting Book*. The pages in the Spanish edition appeared lighter than those in the softcover edition reviewed, which made it somewhat easier to count the objects on some pages. See the review of Pallotta's *The Icky Bug Counting Book*.

Reiss, John J. *Numbers*. New York: Bradbury Press, 1982.

ISBN 0-02-776150-9		$13.95
ISBN 0-689-71121-2	(paper)	3.95

☆ Single concept, PS–2

Number values from 1 through 20 and multiples of 10 through 100 are colorfully illustrated in effective groups with plainly identifiable and relevant objects. Occasionally, confusion arises when the subdivisions of a single object are used to illustrate a number concept; for example, a clover leaf illustrates the concept of 3. The last page, illustrating raindrops, depicts the concept of 1000.

Sis, Peter. *Waving*. New York: Greenwillow Books, 1988.

ISBN 0-688-07159-7		$11.95
ISBN 0-688-07160-0	(library binding)	$11.88

☆ Single concept, PS–2

Mary's mother waved for one taxi, but instead of a taxi, two bicyclists waved at Mary and her mother, and then three boys walking dogs waved at the waving bicyclists. Continuing misinterpretations of who is waving at whom introduce numbers from one through fifteen and their corresponding numerals. The final set comprises fifteen taxi drivers waving at everyone; so Mary and her mother walk home. Color is effectively used to introduce each new set against gray urban street scenes. Watercolor and pen-and-ink illustrations and a simple text tell a humorous, appealing story.

Sloat, Teri. *From One to One Hundred.* New York: Dutton Children's Books, 1991.
See Early Number Concepts: Counting—Place Value

Smith, Maggie. *Counting Our Way to Maine.* New York: Orchard Books, 1995.
ISBN 1-531-06884-6 $14.95
ISBN 1-531-06734-4   (library binding)
☆ Single concept, PS–1

Summer vacation on the Maine coast is the context for *Counting Our Way to Maine;* the numbers from one through twenty provide the structure for sequencing the images. The opening scenes of a family busily packing for their summer vacation—one baby, two dogs, three bicycles—and their trip through busy urban streets—four taxicabs, five smoke stacks—gradually change to serene scenes of eight lush green mountains and glimpses of nine deer. Number provides the link between each scene. At their destination, other images of Maine (and number) include eleven sand castles, twelve lobster traps, fifteen boxes of blueberries, eighteen mosquito bites, and twenty fireflies. Illustrations through watercolor, gouache, and pastel with ink-and-pencil line are pleasing. Large black numerals accompany the text with number word names incorporated in the simple story line. The objects to be counted are often small and scattered in different forms throughout the scene. This is not a criticism, since beginning counting is not the focus of the book. Numbers are the vehicle for tying together the images of Maine and presenting the view that opportunities to count are all around us.

Thornhill, Jan. *The Wildlife 1, 2, 3: A Nature Counting Book.* New York: Simon & Schuster Books for Young Readers, 1989.
ISBN 0-671-67926-0 $16.00
★★ Single concept, PS–2

Elaborate, but controlled, artwork graces the pages of this beautiful, large-sized book of wild animals featured in their native setting; each scene is framed with an intricate motif. An appropriate set of animals can be counted in each scene, and within the border, a second corresponding set has been included. The numeral and its

word name, as well as the name of the animal, are printed under
each framed illustration. The numbers are introduced sequentially
from 1 through 20, then 25, 50, 100, and 1000 are represented. A
number of the sets are difficult to count because of the size of the
number as well as the busy pattern of the scene—the 25 butterflies,
50 flamingoes, 100 penguins, and 1000 tadpoles. As with most
illustrations for representing large numbers, the purpose becomes
developing number sense rather than counting. To show this concept
of increasing size, the final pages display the sets from 1 through 20
and the sets from 25 through 1000 illustrated in triangular forms,
with the 20 fish and the 1000 tadpoles forming the base of each
triangle and the 1 panda and the 25 butterflies forming the opposite
vertex, respectively. A "Nature Notes" section includes detailed
information about each animal and its environment.

Tildes, Phyllis Limbacher. *Counting on Calico.* Watertown, Mass.:
Charlesbridge Publishing, 1995.

ISBN 0-88106-863-2		$14.95
ISBN 0-88106-864-0	(library reinforced)	15.00
ISBN 0-88106-862-4	(paper)	6.95

☆ Single concept, K–4

Willie Whiskers, "squeaker of the house," introduces the reader
to calico cats and counting from 1 through 20. Facts about cats form
the basis of many of the counting tasks: "1 tail / 2 big eyes / 3 black
spots / 4 legs / 5 paw pads / 6 orange spots." Other opportunities to
count include front claws, birds, paper sacks, food bowls, and
whiskers. One reference to number is abstract (nine lives), but shows
nine cats to count. Some of these tasks become more challenging
depending on the size of the number or the type of object to be
counted. For each new cat fact, Willie Whiskers appears beside a
paragraph printed in small type that informs the reader about
characteristics of calico cats and cats in general. The appealing
illustrations are detailed and colored in pastel shades.

Tucker, Sian. *1 2 3 Count with Me.* New York: Little Simon, 1996.

ISBN 0-689-780828-3 $12.95

★★★ Single concept, (PS–2)

Numbers from one through twenty are featured in this well-
designed, lift-the-flap counting book. Each lift-the-flap is cleverly
designed to provide a second opportunity to count. When the rocket's
door is opened, the reader sees one astronaut seated inside.
Underneath the ear (flap) of one of the two elephants are two tiny
mice. Underneath the hood (flap) of one of the five racing cars are
five wrenches. Bold color is effectively used to highlight the large fat
numerals and the simply constructed objects against a contrasting
colored background. Minimal text introduces each number word
name and the sets to be counted. A final surprise is the page

featuring twenty. Twenty presents are hidden under twenty flaps with an invitation to guess what each package contains. Each present relates to one of the nineteen previous pages.

Tudor, Tasha. *One Is One.* New York: Simon & Schuster Children's Books, 1956.

ISBN 0-02-688535-2 (paper) $4.95
ISBN 0-689-71743-1 (paper) 4.95
☆ Single concept, PS–1

Tasha Tudor's distinctive rosy-cheeked children are here for the counting. Charming illustrations and original verse present numbers and their values, one through twenty. The appealing quality of Tudor's artwork makes the book a pleasure to read and reread, although the placement of text and illustrations is sometimes misleading.

Watson, Amy, and **the Abby Aldrich Rockefeller Folk Art Center Staff.** *The Folk Art Counting Book,* Williamsburg, Va.: Colonial Williamsburg Foundation and New York: Harry N. Abrams, Publishers, 1992.

ISBN 0-87935-084-9 $9.95
☆ Single concept, PS (all ages)

Conceptualized by Florence Cassen Mayers, an author of museum alphabet books, this book represents numbers from one through twenty by using selected art from the Abby Aldrich Rockefeller Folk Art Center. Large numerals are artfully presented in a rectangular frame with a graceful folk art border. Each page tells what to count: "count 1 whirligig" beneath a carved wooden statue or "count 2 monkeys" beneath a painting by Henry Church. Some pages include two opportunities for counting and displaying art work; for example, 16 Native Americans can be counted on the watercolor *Indian War Dance* and 16 diamonds on the *Quilt Square—Rollins Family Record.* Some objects are more difficult to count, like those in the painting *Dark Jungle,* where the jungle animals are so small and the painting is so dark. The main purpose of this counting book is to introduce folk art; documentation for all the art is included in the back of the book. The selections are well presented but may lack appeal to intrigue children.

Wood, Jakki. *One Tortoise, Ten Wallabies: A Wildlife Counting Book.* New York: Bradbury Press, 1994.

ISBN 0-02-793393-8 $14.95
★★★ Single concept, 2+

This book presents a truly aesthetic experience as carefully chosen and highly descriptive text is complemented by simple lines and deliciously delicate watercolors. A procession of animals walks, hops, flips, swings, and rollicks across the pages. Large bold numerals and word names describe each new countable set. From "one slow tortoise," "two hopping hares," "three flying ducks," to "fifteen strong, stripy, rough and tumbly tigers," the groups move quietly forward. Then the pattern changes to represent 20, 25, 50, and finally 101. This page, which must be opened out, uses both sides to accommodate the herd of 101 "big, bigger, biggest, small, playful, powerful, plodding elephants." This happy and vibrant celebration of the animal world is a must to read—a most satisfying book!

Counting—
Place Value

The following description of a first-grade classroom in Cedar Falls and the children's responses to *Count and See* by popular author and photographer Tana Hoban should help illustrate the potential that these kinds of books can have in the development of mathematical processes and number concepts:

> The first graders were sitting on the floor around their teacher, Sue Sherwood, as she read to them after lunch. The children verified each set by counting that the number of objects depicted was correct. On one page, they noticed that the counters were grouped in fives; so the children counted by fives to determine the total. On another page, after a child counted to verify that ten counters were in each set, the first graders counted by tens. The teacher asked them, "If we count by tens, we get fifty. How many groups of ten is fifty?"
>
> As part of the lesson, the teacher asked the students to predict the next number after each page. After sharing the pages showing the first fifteen counting numbers, the teacher told them that the pattern had changed. After they discussed the next number, which was twenty, they correctly predicted that the next number was thirty. They continued with their predictions until they predicted that sixty would follow fifty. At this point, the teacher explained that that was a good guess but that the pattern had changed again, and there was only one page left in the book. In unison, the first graders excitedly yelled, "One hundred! One hundred!" They continued with their mathematics lesson because they were currently studying place value.

This group of counting books has many of the characteristics of other counting books that have been mentioned in the "Counting—One to Ten" and the "Counting—Large Numbers" categories. They are grouped into a separate category because they, unlike the others, imply the concept of place value; objects are explicitly arranged in sets of ten to encourage counting by tens. McMillan's *Counting Wildflowers* provides an excellent format for grouping and for developing number sense for the numbers from 1 through 20. Cave's *Out for the Count: A Counting Adventure* uses silhouettes of rows of ten to highlight a number from each decade from 1 through 100. Manushkin's *Walt Disney's 101 Dalmatians: A Counting Book* can be used as a bridge for counting across decades. A sophisticated treatment of the concept of place value is found in Howard's triumphant *I Can Count to a Hundred ... Can You?* This category contains many exceptional books for extending counting skills to beginning place-value concepts.

Anno, Mitsumasa. *Anno's Counting Book.* New York: HarperCollins Children's Books, 1977.

See Early Number Concepts: Counting—Zero

Cave, Kathryn. *Out for the Count: A Counting Adventure.* Illustrated by Chris Riddell. New York: Simon & Schuster Books for Young Readers, 1992.

ISBN 0-671-75591-9 $14.00

★★★ Single concept, PS–3

When Tom couldn't sleep, his father suggested, "Try counting sheep." Tom's wide-eyed imagination takes form as 6 of the sheep from his bedcover come to life and lie down on his bedroom floor. A seventh sheep beckons to him to follow on a wild and woolly adventure. Tom and his stuffed toy rabbit have many a narrow escape as they avoid the dangers of 12 wild wolves, 23 pythons, 36 mountain goats, ..., 97 Bengal tigers. The illustrations and text clearly document Tom's dilemmas as he escapes from one situation into a new one. One number in each decade from 1 through 100 is highlighted. The detailed illustrations show each number in two ways. For example, 54 penguins crowd around Tom and his friend on an icy shore. The 54 penguins are also shown as silhouettes of five rows of 10 and 4 more. This clever design clearly and effectively presents beginning place-value concepts. The inside cover shows a game board with the numbers from 1 through 100. Cave's beautiful fresh and lively illustrations complement the humorous rhyme and should appeal to young readers as they extend their concepts of number. They may find it interesting to discover that all the animals encountered in the story can be found on Tom's bed in the first illustration.

Fleming, Denise. *Count!* New York: Henry Holt & Co., 1992.

ISBN 0-8050-1595-7 $14.95

★★ Single concept, PS–1

Handmade paper by the author is the basis of the stylized artwork in this minimal-text number book. Done in an abstract style, the brilliantly colored illustrations of cranes, giraffes, toucans, lizards, and other lively critters present the numbers from one through ten and then twenty, thirty, forty, and fifty. Each representation includes the appropriate number of creatures and rectangular-shaped counters, the numeral, and the word name. The counters for the multiples of ten are shown in groups of ten. The forty frogs leaping across the page will be counted one by one, whereas the counters are clearly arranged in four groups of ten for ease in counting.

Hoban, Tana. *Count and See.* New York: Macmillan Books for Young Readers, 1972.
ISBN 0-02-744800-2 $14.95
★★★ Single concept, PS–2

Number concepts from 1 through 15, 20, 30, 40, 50, and 100 are introduced. Each spread includes the number symbol, name, and appropriate number of counters on one page, with the opposite page featuring Hoban's distinctive photographs that capture familiar objects from a child's perspective. Multiples of ten are clearly shown as arrays of counters on the left. The objects in the photographs are clustered in sets of five or ten for ease in counting. Counting and its connections to the reader's world are obvious. The presentation is simply done in black and white—with dramatic results.

Howard, Katherine. *I Can Count to One Hundred ... Can You?* Illustrated by Michael J. Smollin. New York: Random Books for Young Readers, 1979.
ISBN 0-394-84090-9 (paper) $2.25
★★★ Multiconcept, PS

Narrated by a busy, cheery mouse, this appealing counting book has many unique features. Each double-page layout introducing the numbers from one through ten has more than one set to count. Each numeral and number word is boldly displayed; each narrative includes the entire word list that the young readers say as they count orally. A summary page of the numbers to ten is followed by the numerals and word names for ten to twenty, complete with counters arranged in bar-graph form. The pattern of decade numbers is effectively illustrated in an example from twenty to thirty. Each number is expressed in a form emphasizing the decade value and the ones value. For example, twenty-three is shown as 20 + 3 = 23 with a set of twenty orange counters and three green counters. The numeral 20 and the 2 in 23 are printed in orange, and the 3s, the ones numerals, are printed in green. Another double-page layout shows a hundreds-chart with each decade printed in a different color; the word name for each new decade number is printed beside the numeral. The design and illustrations provide an excellent opportunity for children to experience counting and number patterns. The concepts of this book could be explored by children of many ages.

Lottridge, Celia Barker. *One Watermelon Seed.* Illustrated by Karen Patkau. New York: Oxford University Press, 1990.

ISBN 0-19-540735-0   (paper) $6.95

★★   Single concept, PS–2 (1–3)

Two children, Max and Josephine, carefully plant their garden from one watermelon seed, two pumpkin seeds, three eggplant seeds, ..., to ten corn seeds. The second section of the book depicts the results of their labor as the children harvest 10 watermelons, 20 pumpkins, 30 eggplants, ..., 100 ears of corn. The initial illustrations of the spring landscape are dominated by the brown earth, which gives way to the colors depicting the harvest of orange pumpkins, red and green ripe melons, and deep purple eggplants. Each seed results in a tenfold harvest. A final surprise is that although the 10 seeds of corn yield 100 ears of corn, they would yield thousands of popcorn kernels. Some of the examples for the multiples of 10 are easier to count than others, which should help the reader realize the power of grouping by 10. The 80 string beans are arranged in 8 groups of 10, but the basket of 60 blueberries and the tray of 70 strawberries are more difficult to count. The word name for each number concept is contained in the simple text; an appropriate bold black numeral accompanies each grouping.

McMillan, Bruce. *Counting Wildflowers.* New York: Lothrop, Lee & Shepard Books, 1986.

ISBN 0-688-02859-4 $16.00
ISBN 0-688-02860-8   (library binding)   15.93
ISBN 0-688-14027-0   (paper, Mulberry)   4.95

★★★   Single concept, PS–1

Daylilies, sundrops, orange hawkweeds, and mullein pinks are some of the wildflowers whose beauty is captured in McMillan's color photographs. Across the bottom panel of the first ten pages is a row of ten counters. Two purple counters are in one-to-one correspondence with the two purple spiderworts pictured; the remaining eight counters are green. This pattern continues until eleven red Maltese crosses are matched with one row of red counters plus one more red counter in a second row of ten counters. Each representation clearly shows beginning place value as groups of ten. The green counters also effectively illustrate how many more are needed to obtain ten or another ten. The only text is the name of each type of flower and the appropriate numeral and number word for each number. Additional information about when and where to find these wildflowers is included in the final pages.

Manushkin, Fran. *Walt Disney's 101 Dalmatians: A Counting Book.* Illustrated by Russell Hicks. New York: Disney Press, 1991.

ISBN 1-56282-012-5		$9.95
ISBN 1-56282-032-X	(library binding)	9.89
ISBN 1-56282-324-8	(paper)	4.95

★★ Single concept, PS–K

Roger and Anita's walk with their dalmatians, Pongo and Perdita and their puppies, is interrupted when the puppies scatter after being startled by a fire truck. The remainder of the story depicts the search for the puppies, which are found in groups of 10. It is a challenge to find all 10 puppies scattered throughout each new scene. For each layout, the reader is asked to keep an accumulative count: "Find 10 more puppies and count them from 21 to 30 ... from 31 to 40...." Concern is expressed when 99 puppies are found—but there should be 101 dalmatians. The reader is asked to solve the problem. The format clearly links counting to place-value ideas.

———. *One hundred one dalmatas: Un libro para contar.* New York: Disney Press, 1994.

ISBN 1-56282-697-2	(library binding)	$13.89
ISBN 1-56282-568-2	(paper)	5.95

★★ Single concept, PS–K

Originally published as *Walt Disney's 101 Dalmatians: A Counting Book,* this Spanish version has been more recently released. See the review of Manushkin's *Walt Disney's 101 Dalmatians: A Counting Book.*

Sloat, Teri. *From One to One Hundred.* New York: Dutton Children's Books, 1991.

ISBN 0-525-44764-4	$14.99

☆ Single concept, PS–2

Counting opportunities for numbers from 1 through 10 and then for multiples of 10 from 20 to 100 abound in this no-text number book. Each busy page provides many counting opportunities; for example, in one scene depicting a bathroom, the reader can count four seagulls, four dolphins swimming in the bathtub, four towel rings, four towels (not necessarily on the towel rings), four periscopes, and four puddles on the tile floor. Large water pipes are carefully drawn to form a large numeral 4. Along the bottom border of each page, pictures of the objects to be counted are given as well as the appropriate numerals and word names. In each fantasy scene some objects are grouped so that they are easier to count than other randomly arranged objects. The Academy of Acrobatics is the theme for a two-page spread that has forty acrobats scattered throughout the scene. Forty stars are arranged to form the numeral 40, which is bordered by two rows of twenty stagelights. Five floors of two high-rises can be seen; each floor has windows arranged in rows of four. Four tightropes strung between the high-rises also serve as clotheslines as ten towels hang from each. Each of the detailed, pastel illustrations should capture the reader's imagination. Children from many ethnic backgrounds are depicted throughout the book.

Sugita, Yutaka. *Goodnight, 1, 2, 3.* New York: Scroll Press, 1971.

See Early Number Concepts: Counting—One to Ten

Counting—
Bar Graphs

The end pages that appear in the books selected for this category may have been designed to be an introduction to what is to come and a summary of what has been. The graphs clearly show "one more than" or "one less than" relationships between consecutive numbers. But more than that, they lend themselves to an introduction to summarizing and documenting information through simple pictographs. This format could encourage the young reader to use graphing activities to represent consecutive numbers or other data, since collecting, representing, and analyzing data are important topics for students of all ages.

Calmenson, Stephanie. *Ten Furry Monsters.* Illustrated by Maxie Chambliss. New York: Parents Magazine Press, 1984.
See Early Number Concepts: Counting—Operations

Carle, Eric. *Rooster's Off to See the World.* Natick, Mass.: Picture Book Studio, 1971.
ISBN 0-88707-042-1 $15.95
ISBN 0-88708-178-9   (paper)   4.95
★★★ Single concept, PS+
    Rooster sets off to see the world and is joined by two cats, three dogs, four turtles, and five fish. As readers travel with Rooster and his companions, they have a graphic opportunity to count forward and backward as Rooster is gradually deserted by the sets of animals who had joined him on his journey. The book is beautifully illustrated in the vibrantly colorful collage technique so typical of Eric Carle's work. Carle reinforces the mathematical concepts with tiny silhouetted diagrams on the top right-hand side of every other page. These diagrams could be used as an introduction to bar graphs.

Coats, Laura Jane. *Ten Little Animals.* New York: Macmillan, 1990.
See Early Number Concepts: Chants, Songs, and Rhymes

Crews, Donald. *Ten Black Dots*. New York: Greenwillow Books, 1986.

ISBN 0-688-06067-6		$16.00
ISBN 0-688-06068-4	(library binding)	15.93
ISBN 0-688-13574-9	(paper, Mulberry)	4.95

☆ Single concept, PS–3

Colorful drawings contain big black dots that represent the cardinal numbers from one through ten. The big black dots show up in such commonplace ways as three dots on a snowman's face, four knobs on a radio, and five buttons on a shirt. The simple rhyming text introduces the numeral, the word name, and two examples for counting each number. At the end of the book, a chart of the numbers from one through ten clearly shows the "one more" concept. This appealing book for young children is often used to motivate them to make their own "Ten Black Dots" book.

————. *Diez puntos negros*. New York: Scholastic, 1995.

ISBN 0-590-48657-8   (paper) $4.95

Gardner, Beau. *Can You Imagine ...? A Counting Book*. Messapequa, N.Y.: BGA, 1995.

ISBN 0-9639898-2-0   (paper)

★★★ Single concept, PS–2

Can you imagine one whale wearing a veil or twelve swans twirling batons? Zany humor, reflected in strong, vibrant illustrations and rhyming text, presents the numbers from one through twelve. The funny antics of the animals are accompanied by the appropriate numerals and number words. At the end of the book, the number concepts are reviewed effectively by a chart of silhouetted forms of the animals featured so imaginatively throughout the text. This chart clearly illustrates the "one more than" concept.

Koch, Michelle. *Just One More*. New York: Greenwillow Books, 1989.

ISBN 0-688-08127-4 $11.95
ISBN 0-688-08128-2   (library binding)

★★ Single concept, PS+

Pairs of singular and plural forms—goose and geese, child and children, foot and feet, knife and knives, moose and moose—label each childlike watercolor of the appropriate objects. At first glance, this book appears to be simply about the plural forms of irregular nouns. On closer examination, the reader will notice that the number of objects that illustrates each plural form is always one more than the previous set. Sequentially, there are two children, three geese, four moose, five leaves, ..., fifteen sheep. On the final page, each set from one through fifteen is arranged and labeled in a vertical row; this bar-graph arrangement effectively shows that each subsequent set is one more. No number word names are noted; the numerals are introduced on only the closing page.

Kosowsky, Cindy. *Wordless Counting Book.* Bridgeport, Conn.: Greene Bark Press, 1992.
See Early Number Concepts: Counting—Zero

MacKinnon, Debbie. *How Many?* Illustrated by Anthea Sieveking. New York: Dial Books for Young Readers, 1993.
ISBN 0-8037-1253-7 $10.99
★★★ Single concept, PS–K

Color photographs of children at play introduce numbers and numerals from one through ten. Each number is introduced with a question such as "How many teddy bears is Faye playing with?" and a close-up of a child playing with teddy bears arranged in a basket or on a chair. The answer of seven, as well as the appropriate numeral and word name, can be found on the opposite page, where the same seven teddy bears have been rearranged for ease in counting. This page is framed with a brightly colored border liberally splashed with small white numerals. All the children and objects have been photographed with a white background, which removes unnecessary clutter—an asset for young children. Children from different ethnic backgrounds have been included to introduce each example. The end pages summarize the numbers from 1 through 10 with the bright yellow numerals boldly announcing the counters, arranged in a bar-graph format, that are introduced in the book.

Nikola-Lisa, W. *One, Two, Three Thanksgiving!* Illustrated by Robin Kramer. Morton Grove, Ill.: Albert Whitman & Co., 1991.
ISBN 0-8075-6109-6 $14.95
ISBN 0-8075-6110-X   5.95
★★★ Single concept, PS–1

Thanksgiving dinner preparations provide the context for counting from one through ten as Papa stuffs one turkey, Mama bakes two pies, little sister measures three cups of cranberries, …, and Grandma lights ten candles. Lined up above each fresh watercolor illustration are the appropriate numeral and the objects to be counted. Each scene of the family's activities clearly shows the objects to be counted; the appropriate numeral is also incorporated into the scene's design. After all ten family members share their meal, the numbers from ten through one are recounted in the simple text and corresponding illustrations as the necessary cleanup begins, from Papa stacking ten empty plates to Grandma breaking into one big smile. The opening and closing pages show the numerals and the appropriate number of objects arranged in a chart from 1 through 10 and 10 through 1, which clearly shows the "one more than or one less than" concept between adjacent numbers. The total experience is satisfying.

Ormerod, Jan. *Young Joe.* New York: Lothrop, Lee & Shepard Books, 1986.

ISBN 0-688-04210-4 $4.95

☆ Single concept, PS

Joe, a young African American child, counts off sets of animals from one fish through ten piglets by holding up the appropriate number of fingers for each set. The illustrations simply and clearly complement the minimal text that introduces the word names from one through ten. All the animals are illustrated in pictograph format in the book's endpapers. The book's beginning pages show the "one more than" concept of sequential counting because the animals are arranged in columns and are labeled by the numerals from one through ten. The last pictograph depicts the same columns of animals arranged from ten through one.

Price, Marjorie. *1 2 3 What Do You See?: A Book of Numbers, Colors, and Shapes.* Stamford, Conn.: Longmeadow Press, 1995.

ISBN 0-681-10175-X $7.95

★★ Multiconcept, PS–2

Bold colors on deep purple pages provide rich background for an unusual counting experience from one through ten. A large white numeral appears on each left-hand page; the word name follows in the description of colors and shape, for example, "Three zany zig-zags jump up and down. / What color are they?" Below are three bright yellow zigzags for the reader to count. The opposite page is filled by a huge stylized numeral composed of many colors; the reader is asked to find the yellow zigzag in the numeral shape. Since each shape is associated with a specific color, the reader has two clues for finding it. The shapes that are introduced are informal, such as half circles, wiggle waves, stars, arrows, and curlicues. Throughout the book, the objects are clearly illustrated, which makes them easy to count. A final page shows all numerals from 1 through 10 with an appropriate number of shapes.

Tryon, Leslie. *One Gaping Wide-Mouthed Hopping Frog.* New York: Atheneum, 1993.

See Early Number Concepts: Counting—One to Ten

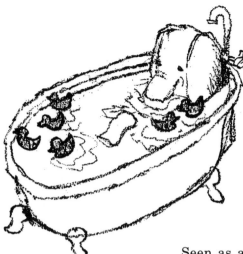

Counting—
Partitions

Seen as a beginning step toward understanding the many combinations in which number sets can be represented, these books have been set apart because they contain an introduction to partitions. Facility with partitions is essential for number sense and mental computation. Part of the set of six can be found hidden in the grandfather clock in Dodds's *Someone Is Hiding*. *Anno's Counting House* is an excellent introduction to partitions of ten. Merriam's *Twelve Ways to Get Eleven* is a natural springboard for exploring partitions of other numbers and composing one's own book. In Johnson and Johnson's *One Prickly Porcupine* and the books cross-referenced in this category, the countable objects can be found arranged in various combinations.

Anno, Mitsumasa. *Anno's Counting House.* New York: Philomel Books, 1982.

ISBN 0-399-20896-8 $16.95

★★★ Single concept, PS–3

One by one, ten children move from their current home to the house next door. Readers will be intrigued by this simple plot, and the unique design puts this book in a class of its own. The interior of the house being vacated is always on a left-hand page; the house the children are moving into is shown on a right-hand page. The pages are separated by a divider page whose sides show the stately exterior of each house. Thus when the readers open the book, they see the interior of the old house and the exterior of the new house. When the divider page is turned, the interior of the new house and the exterior of the former house appear. Window cutouts in the divider pages give readers a glimpse of some of the children in the four levels of the house, but the readers cannot verify the number of children in the house until they turn the page. The format offers a fascinating approach to the partitions of ten. The house interiors are illustrated in detailed black-and-white drawings; the ten children are in color. An introductory note to readers explains the use of the book; a section at the back gives parents and other adults more suggestions.

Audry-Iljic, Francoise, and **Thierry Courtin.** *My First Numbers.* Adapted by Judith Herbst. Hauppauge, N.Y.: Barron's Educational Series, 1994.

See Early Number Concepts: Counting—Other Number Concepts

Dodds, Dayle Ann. *Someone Is Hiding.* Illustrated by Jerry Smath. New York: Little Simon, 1994.

ISBN 0-671-75542-0 $8.95

★★ Single concept, PS–2

The big grandfather clock at the back of the room says noon; the happy birthday sign and the colorful decorations declare that Bunny is about to celebrate his bithday as an assortment of six delightfully clad, personality-plus animal guests appear at his door with their gifts. By 1 o'clock, the friends are playing hide-and-seek and one is missing. Despite a fun-filled search, they have no luck finding their friend. As each hour progresses, the pattern continues. Alternate pages allow the reader to peek under the flip-up door of the grandfather clock to see where the lost friends can be found. But their location is a secret to the rest of the searching animals. By 6 o'clock, the only animal left in view is the birthday bunny. When he hears a giggle, he opens the clock door and his friends tumble at his feet. A consistent thread throughout the book is the tin of sardines, the gift from the little crocodile. At first it is closed. As 1 character is hidden in the clock, the sardine tin is opened to show one sardine, as 2 are hidden, two sardines are shown, and so forth. Numerals rather than word names are used in the text. Mathematics concepts embedded in the story line include counting, beginning addition and subtraction, and the passage of time by hours. Partitions of seven are implied, since the number of animals hiding in the clock complements the number seeking the lost friends. The illustrations demand and deserve a serious look. The expressions on these captivating characters' faces, their clothing, and their antics are all hilarious.

Grande Tabor, Nancy Maria. *Cincuenta en la Cebra: Contando con los Animales.* / *Fifty on the Zebra: Counting with the Animals.* Watertown, Mass.: Charlesbridge Publishing, 1994.

See Early Number Concepts: Counting—Other Number Concepts

Johnson, Odette, and **Bruce Johnson.** *One Prickly Porcupine.* New York: Oxford University Press, 1991.

ISBN 0-19-540834-9 $13.95

★★ Single concept, PS+ (PS–1)

Brightly colored illustrations, composed of clay and plasticine, dominate each busy, detailed illustration that introduces the numbers from one through twenty. On the page depicting fourteen, the reader has the option of counting fourteen ladybugs lined up on the numeral 14, fourteen butterflies hovering over fourteen presents, fourteen frogs riding in three baskets held up by fourteen balloons, fourteen buttons on the giant's coat or the fourteen holes in the giant's belt, as well as the fourteen spiders highlighted in the page's caption, "Fourteen splendid spiders spinning on some hips." The beginning counter may find the illustrations on some pages challenging. For example, the four busy beavers are almost lost in the competing detail. Others may rise to the challenge as they note that each of the four beavers wears a hat and has four fingers on each foot; each of four bees has four stripes, a pair of wings, and a hat. In each illustration a large, bright numeral announces each number.

Long, Lynette. *Domino Addition.* Watertown, Mass.: Charlesbridge Publishing, 1996.
See Number—Extensions and Connections: Operations

McMillan, Bruce. *Jelly Beans for Sale.* New York: Scholastic Press, 1996.
See Measurement: Money

Merriam, Eve. *Twelve Ways to get Eleven.* Illustrated by Bernie Karlin. New York: Simon & Schuster, 1993.
See Number—Extensions and Connections: Number Concepts

Wahl, John, and **Stacey Wahl.** *I Can Count the Petals of a Flower.* Reston, Va.: National Council of Teachers of Mathematics, 1985.
See Early Number Concepts: Counting—Other Number Concepts

Counting—
Operations

Although Betsy Lewin's *Cat Count* is now out of print, the description below shows how one group of first graders were challenged with the concept of "how many in all."

As Sue Sherwood read *Cat Count* to her first graders, one child observed that the page said "plus four" rather than "four." The teacher asked them to guess what that might mean, and the children suggested that maybe they would add all the cats. Their predictions were confirmed; the illustration after "plus five" showed the accumulation of all the cats. As they continued through the book, the teacher asked them to estimate how many total cats there would be after they were introduced to the set of ten cats. They were discussing the reasonableness of different guesses like 100 and 1000 when one first grader announced, "There are fifty-five cats!" The child explained his reasoning by quickly holding up one finger, then two fingers, then three fingers, and so on, while rapidly counting aloud. To help others understand his thinking, the teacher asked ten children to stand in a line and hold up from one to ten fingers. As a class they counted the fingers and verified the child's conjecture.

Still within the broad category of counting books and still containing the many mathematical virtues already discussed as characteristic of this genre, the books annotated below have been assigned to this category because of their potential for connecting the counting function with concepts on a more advanced level of mathematical understanding. Addition or subtraction concepts are implied in some books, since the same set is increased or decreased by similar objects. Examples include Gretz's *Teddy Bears 1 to 10* (the teddy bears appear one by one) and Wise's *Ten Sly Piranhas: A Counting Story in Reverse* (the piranhas disappear one by one). Counting on and counting back are both included in Walsh's *Mouse Count*. Subtracting one and comparing two consecutive numbers occupy center stage in Rocklin's *Musical Chairs and Dancing Bears*. In

Calmenson's *Dinner at the Panda Palace,* the accumulative count of the diners is fifty-five; this result can lead to developing strategies for adding the numbers from one to ten.

Arnold, Tedd. *Five Ugly Monsters.* New York: Scholastic, 1995.
See Early Number Concepts: Chants, Songs, and Rhymes

Audry-Iljic, Francoise and **Thierry Courtin.** *My First Numbers.* Adapted by Judith Herbst. Hauppauge, N.Y.: Barron's Educational Series, 1994.
See Early Number Concepts: Counting—Other Number Concepts

Beck, Ian. *Five Little Ducks.* New York: Henry Holt & Co., 1992.
ISBN 0-8050 2525-1 $14.95
☆ Single concept, PS–2
"Five little ducks went swimming one day, / Over the hills and far away." As the repetitive rhyme continues, the little ducks disappear one by one. The threatening presence of a fox leads the reader to anticipate the worst. But, at the story's end the last little duck finds the others and reunites them with mother duck. The story line and the illustrations introduce beginning concepts of subtraction. For ease in counting, the little ducks are clearly and simply depicted in the pleasing illustrations that are done primarily in shades of blue, green, and yellow. Very young children will enjoy reading along with the repetitive verse.

Becker, John. *Seven Little Rabbits,* 2d ed. Illustrated by Barbara Cooney. New York: Walker & Co., 1994.
ISBN 0-8027-8311-2 $5.95
ISBN 0-590-63196-9 (Scholastic, 1985) 6.95
ISBN 0-590-44849-8 (paper, Scholastic, 1991) 3.95
★★ Single concept, PS–1
Repetitive, rhyming verse encourages children to count backward from seven to one. Seven rabbits leave home to visit old friend toad. As each rabbit tires and journeys back to mole's house, one more rabbit is tucked into mole's bed. The story ends with all seven rabbits tucked into mole's bed, mole sleeping in a chair, and toad wondering where his friends are. The whimsical illustrations enhance the text with additional humor. The story also introduces beginning subtraction concepts as "one less than" the previous set.

Bucknall, Caroline. *One Bear All Alone.* New York: Dial Books for Young Readers, 1985.
ISBN 0-8037-0238-8 $9.95
ISBN 0-8037-0645-6 (paper) 4.95
☆ Single concept, PS–2

Clever illustrations of each numeral from 1 through 10 dominate the left-hand page. The opposite page is richly illustrated in warm and winsome scenes depicting the activities of an appropriate number of appealing bears. Each number and scene is introduced simply through rhyme. The final layout depicts all ten bears romping in a row, with the numerals from 1 through 10 emblazoned on their sweatshirts.

Bursik, Rose. *Zoë's Sheep.* New York: Henry Holt & Co., 1994.

ISBN 0-8050-2530-8 $14.95

★★ Single concept, PS–1

When Zoë protests that she is not tired, her father, as he tucks her into bed, suggests that she count sheep to fall asleep. Rhyming verses introduce, one by one, ten sheep who dramatically enter through the open bedroom window. "Sheep number one liked to dance, number two was fond of plants." This nonsense rhyme is complemented by whimsical illustrations of energetic, talented, and comical sheep. Sheep number one continues to dance on top of the furniture while sheep number two happily eats one large plant and starts on the potted flowers. Each new visitor is a unique surprise, and the reader will continue to watch for the humorous antics of each sheep in the remaining scenes. The deep brilliant blues of the bedroom walls and the sky outside are a lovely background for the soft touches of color in Zoë's room and the white-gray, fluffy-looking sheep. The young child should enjoy this opportunity for counting and finding each unique sheep. In the last scene, ten sheep are not seen because the tenth sheep scares the others away. Number names from one through ten are embedded in the verse.

Calmenson, Stephanie. *Dinner at the Panda Palace.* Illustrated by Nadine Bernard Westcott. New York: Scholastic, 1991.

ISBN 0-590-62389-3

★★★ Single concept, (PS–3)

Calmenson's rhyming verse and Westcott's pleasing illustrations combine for a cumulative count of the numbers from one through ten, the number of guests at the Panda Palace. Mr. Panda's first guest, an elephant, requests, "I'm enormously hungry. / My bag weighs a ton. / I would like to sit down. / Have you a table for one?" Mr. Panda greets and seats each group of diners, from one hungry elephant, two royal lions, three pigs, and four proud peacocks to mother hen and her nine chicks who fill a table for ten. A crowded scene shows the ten occupied tables; "The waiters moved fast. / Feeding fifty-five diners / Was no easy task." One more guest, a tiny mouse, arrives; gracious Mr. Panda ponders the full house but resolves the dilemma by sitting the mouse on a trunk beside the elephant. The illustrations are lively and humorous: the pigs are eating corn; the bears have honeycombs; Mrs. Hen and her chicks are perched on their chairs. As the dinner closes, various guests are seen carrying bags with extras inside. This book provides a delightful opportunity to introduce the reader to the challenge of adding the numbers from one through ten.

————. *Ten Furry Monsters.* Illustrated by Maxie Chambliss. New York: Parents Magazine Press, 1984.

ISBN 0-8193-1128-6 $5.95

★★ Single concept, PS–3

Gaily colored pictures of ten monster children, each with a numeral emblazoned on the front and back of its shirt, accompany a cheerful story. Subtraction situations occur in the rhyming text as the children disappear from the pages. The illustrations depict the monsters that have "disappeared" as partially hidden. The reader is to count the remaining monsters to verify how many are left. When Mother Monster returns to fetch them from the park, she declares, "I know you're all here somewhere, so I will count to ten." As she counts, they reappear in sequential order. All ten monsters are also shown on the flyleaves in an effective pictograph from 10 to 0 and 0 to 10 that clearly shows "one less than" and "one more than" concepts of counting.

Clarke, Gus. *Ten Green Monsters.* New York: Golden Books, 1993.
See Early Number Concepts: Chants, Songs, and Rhymes

Coats, Laura Jane. *Ten Little Animals.* New York: Macmillan, 1990.
See Early Number Concepts: Chants, Songs, and Rhymes

Dale, Penny. *Ten out of Bed.* Illustrated by Frances Lloyd. Cambridge, Mass.: Candlewick Press, 1993.

ISBN 1-56402-322-2 $14.95

★★ Single concept, PS+ (PS–1)

Lloyd's stunning illustrations combine color, warmth, action, imagination, and detail to complement this wonderful story inspired by the traditional song *Ten in the Bed.* In this story there are definitely ten rambunctious characters out of bed. Not until they have played trains, beaches, acting, pirates, dancing, ghosts, flying, camping, and monsters is the little boy who creates this fantasy tired enough to join his stuffed animals in bed. Tucked in opposite corners in each rich, double-paged illustration are two smaller illustrations. One page shows the little one and six of his stuffed-animal friends in a pirate ship and a merchant ship battling on the ocean. In this example, the box in the top left shows Bear, who had suggested, "Let's play PIRATES!"; in the box in the bottom right, Bear is depicted deep in the sleep of satisfied exhaustion with the caption, "So seven played pirates until Bear fell asleep." In most of the illustrations, the players can be found in different-sized groups, such as four pirates and three others. Partitions of ten are not as evident, since each time one of the characters is shown in the main scene, it is shown again in the lower right-hand corner asleep with the other friends. The text is rhythmic and predictable. This is a "read again and again" book that will become a favorite for enticing a reluctant sleeper to bed.

de Regniers, Beatrice Schenk. *So Many Cats!* Illustrated by Ellen Weiss. New York: Clarion Books, 1988.

ISBN 0-89919-700-0   (paper) $6.95

★★ Single concept, PS–3

Amusing pastel illustrations complement the delightful story line telling how a household grows from an only cat to twelve cats. Each new set of unique arrivals is introduced through irregular verse and appealing illustrations. The reader is asked periodically to count to verify the total number of accumulating cats as they are named and shown in typical feline antics. The charming story presents an opportunity to introduce addition concepts.

Dunbar, Joyce. *Ten Little Mice.* Illustrated by Maria Majewska. San Diego, Calif.: Gulliver Books, Harcourt Brace & Co., 1990.

ISBN 0-15-200601-X $14.00
ISBN 0-15-200770-9   (paper)   5.00

★★ Single concept, PS–1

Majewska's intricate, lavish illustrations of mice and other creatures in their natural habitats are stunning and most realistic. The first double page introduces the ten little mice featured in this narrative rhyme. Nine are shown on the left-hand page; the tenth mouse is seen on the opposite page announcing, "I'm going-going home to my cozy nest." The format continues as one by one the number of mice on each successive left-hand page becomes smaller. The accompanying verse includes the number of the remaining group but always ends with one little mouse leaving to go home to the cozy nest. The final page shows all ten mice in their nest after the last one catches a glimpse of a thunderstorm and "hurry-hurried home to his cozy nest." This simple lesson in subtraction will interest the young reader who will enjoy the action of the little mice and their adventures as they scurry, run, rush, slip, and stagger to their cozy nest.

Ehlert, Lois. *Fish Eyes: A Book You Can Count On.* Orlando, Fla.: Harcourt Brace Jovanovich, 1990.

ISBN 0-15-228050-2 $14.95
ISBN 0-15-228051-0   (paper)   4.95

☆ Single concept, (PS–1)

Bold bright colors sharply define stylized fish as one lone black fish swims through each set outlined against the blue background of the sea. From one green fish to ten darting fish, sets from one through ten are represented and accompanied by the appropriate numeral. Ehlert creates more than the typical counting book, since the reader is encouraged to include the one black fish in the total count. On the page showing five spotted fish, an added line states, "5 spotted fish plus me (the black fish) makes 6."

Enderle, Judith Ross, and **Stephanie Gordon Tessler.** *Six Creepy Sheep.* Illustrated by John O'Brien. Mankato, Minn.: Creative Editions, 1992.
ISBN 1-56397-092-9 $12.95
ISBN 0-14-054994-3 (paper) 4.99
★★ Single concept, PS–1

A countdown tale like its counterpart *Six Sleepy Sheep,* this hilarious story has a much more appealing text. Using lead-on sentences to entice the reader to turn the page and to emphasize the mystery of Halloween night, the narrator follows six unlikely trick-or-treaters who are frightened, one by one, and turn tail to go home. The alliterative text introduces a flock of fairies, a passel of pirates, a herd of hobos, a gaggle of goblins, and a warren of witches. The surprise ending occurs when the last sheep discovers a huge Halloween party—a reader who looks carefully will find that all his friends are already there. The rhythmic quality of the text and the repeated patterns in its development make this story ideal for repeated read-alouds in an early-childhood classroom. Number words are embedded in the text. The countdown activity becomes part of the predicability of the story, and children will enjoy the challenge. The pen-and-ink drawings, rich in detail, reflect the eerie Halloween atmosphere.

———. *Six Snowy Sheep.* Illustrated by John O'Brien. Honesdale, Pa.: Boyds Mills Press, 1992.
ISBN 1-56397-138-0 $14.95
ISBN 0-14-055704-0 (paper) 4.99
☆ Single concept, PS–1

"Six snowy sheep, snug in woolly warm fleece ..." frolic in the snow. In frollicking, slapstick style they test the six presents left under the tree that happy Christmas morning: a sled, skis, ice skates, a saucer, snowshoes, and a shovel. But, swoosh, schmoosh, squoosh—one by one, in countdown style, they land in a deep snowbank. Only the sixth sheep with the sixth gift can help them as he shovels five shakey, shivery, soggy sheep out of the snow. What could be more satisfying than the final picture in which O'Brien, in his inimitable zany style, tucks six contented sheep, with characteristic polka-dot tie, spectacles, stocking cap, and pointy hat, under a warm blanket on a couch in front of a glowing stove. Guess what? Each has a cup of chocolate! AAH! The story is much more satisfying than the quality of the mathematical focus.

Felix, Monique. *The Numbers.* Mankato, Minn.: Creative Editions, 1993.
ISBN 1-56846-001-5 $7.95
★★ Single concept, PS–2

A brown mouse peers through an irregularly shaped cutout on the cover of this square-shaped book that introduces the numbers from one through ten. Clever use of shadows and drawings throughout the book depict mice tearing and cutting out numeral shapes from each stark white page. A beginning page shows a mouse busily tearing the shape of the numeral 1. Both the numeral and its outline from the "torn" page are depicted. On the next page a second mouse appears, looking through an outline from which the numeral 2 seems to be torn from the page. The numeral 2 is stacked on the numeral 1. This pattern continues with the appropriate mice appearing on each page, busily stacking one numeral on another. Two surprises appear at the end of the book—two numerals are torn to represent 10 and a furry paw reaches through the page as the ten mice scramble. Felix's clever design and delicate illustrations are a delight. This format is consistent across her series of Mouse Books.

Gerstein, Mordicai. *Roll Over!* New York: Crown Publishers, 1984.
See Early Number Concepts: Chants, Songs, and Rhymes

Gordon, Jeffie Ross. *Six Sleepy Sheep.* Illustrated by John O'Brien. Honesdale, Pa.: Caroline House, 1991.
ISBN 1-878093-06-1 $12.95
ISBN 0-14-054848-3   (paper)   4.99
Single concept, PS–1

Wacky text and playful illustrations combine in a lighthearted story about six slumbering sheep who wake up when one of them snores. Humans may count sheep to get to sleep, but sheep skip in circles, slurp celery soup, and tell spooky stories until, one by one, they get back to sleep. The last sheep counts to 776 before joining the other slumbering sheep. This countdown story has little to add to the numerous books of its type beyond the zany illustrations and the sheer nonsense of the story.

Gretz, Susanna. *Teddy Bears 1 to 10.* New York: Four Winds Press, 1986.
ISBN 0-02-738140-4   (library binding) $13.95
★★ Single concept, PS–K

Gretz's familiar, lovable teddy bears are the vehicle for introducing the numbers from one through ten. Comical illustrations clearly and simply present the number concepts while showing an enjoyable sequential documentation of events. The text appears on one page; the opposite page colorfully depicts one teddy bear, two old teddy bears, three dirty old teddy bears, and so on. The appealing teddy bears increase by one with each step of a washing, cleaning, and drying process, before ten teddy bears are home for tea.

Henley, Claire. *Joe's Pool.* New York: Hyperion Books for Children, 1994.
ISBN 1-56282-431-7 $12.95
ISBN 1-56282-432-5 (library binding) 12.89
Single concept, PS–1

Eye-popping illustrations in primary colors on a blue background overpower this simple story of a hot summer day. Joe's carefully planned quiet time of sitting in his wading pool is shattered as a friend, a dog, the delivery man, the mail woman, and other overheated people join him in his pool to cool down. The number words in this add-on story are embedded within the simple text but are highlighted with uppercase letters. A large numeral in the top right-hand corner also reflects the growing set. Anticipation builds as the ten characters enter the pool but falls flat with a confusing, rather than satisfactory or humorous, ending. Individuals of different ages and ethnic origins are reflected in the illustrations.

Hunt, Jonathan, and **Lisa Hunt.** *One Is a Mouse.* New York: Macmillan Books for Young Readers, 1995.
ISBN 0-02-745781-8 $13.00
☆ Single concept, PS–1

The spotlight is on as the players from one through ten enter the center stage. "One is a mouse that carried a rose" and "Two is a bear that twirled on its toes" illustrate the rhyming pattern that introduces each new character. Each winsome performer assists in forming a precarious pyramid with the little mouse as the unlikely base. The accumulating pile continues to grow; the mouse confidently holds the group upright until the tenth character, a ladybug, appears. The reader will anticipate the predictable ending as the pyramid crumbles and the animals fall down. The humorous illustrations reflect the playfulness of this simple, zany story. The appropriate numeral is found in the upper right corner of each double page.

Jackson, Woody. *Counting Cows.* San Diego: Harcourt Brace & Co., 1995.
See Early Number Concepts: Counting—Zero

Lee, J. Douglas. *Animal 1*2*3!* Illustrated by Elphin Lloyd-Jones. Milwaukee, Wis.: Gareth Stevens, 1985.
See Early Number Concepts: Counting—Large Numbers

McCarthy, Bobette. *Ten Little Hippos.* New York: Bradbury Press, 1992.

ISBN 0-02-765445-1 $13.95

★★ Single concept, PS–2

The incongruous title of *Ten Little Hippos* is just the first hint of the slapstick humor in this lively countdown book. The unlikely setting for all the action is the stage on which a diminishing number of hippos perform dance routines and skits for the reader's entertainment. The simple rhyming text parallels the action: "Ten little hippos dancing in a line, / One falls away. Now there are...." On the next page the numeral 9 appears on an easel announcing the next act. The hippos' costuming is delightful; their facial expressions are serene. The incongruity of the entire situation contributes to the fun. All ten hippos appear for a final bow and a final count, with the word names from one to ten printed across the bottom of the page. Both end pages are the same but demand examination, since each member of the talented troupe is introduced by his or her eclectic name. Interested readers can reread the story to find out who left first, second, third, and so on. This delightful book should captivate young children.

Moerbeck, Kees, and **Carla Dijs.** *Six Brave Explorers.* Los Angeles, Calif.: Price Stern Sloan, 1988.

See Number—Extensions and Connections: Operations

Nightingale, Sandy. *Pink Pigs Aplenty.* San Diego, Calif.: Harcourt Brace Jovanovich, 1992.

ISBN 0-15-261882-1 $15.00

★★ Single concept, PS–1

Intricate, whimsical illustrations are one of the strengths of this counting book. An opening challenge is to count how many pink pigs dressed in humorous disguises are shown in a busy circus scene. The book explores the numbers from one through ten, which are sequentially introduced through alliterative text and thoughtful illustrations. For example, "Four pink pigs on a prancing pantomime pony" is printed below a design of four pink daisies next to a pink numeral 4, outlined with a green border. At the bottom of the page, the three pigs seen on previous pages are joined by one more pig. On the opposite page are four pigs balancing on a pantomime pony. All the pages are framed with a colored border with a background of pink. The humorous ending is the collapse of the ten-piggy pyramid. "Oops! Perhaps they need more practice" is a final challenge as the reader sees ten spectator pigs on one page and nine performers on the opposite page. The nine performers on this very last page will be counted and recounted until the reader realizes that the missing pig is on top of the pyramid and has yet to fall.

O'Keefe, Susan Heyboer. *One Hungry Monster: A Counting Book in Rhyme.* Illustrated by Lynn Munsinger. Boston: Little, Brown & Co., 1992.
See Early Number Concepts: Chants, Songs, and Rhymes

Pinczes, Elinor J. *Arctic Fives Arrive.* Illustrated by Holly Berry. Boston: Houghton Mifflin Co., 1996.
See Early Number Concepts: Counting—Other Number Concepts

Pomerantz, Charlotte. *One Duck, Another Duck.* Illustrated by Jose Aruego and Ariane Dewey. New York: Greenwillow Books, 1984.
ISBN 0-688-03744-5 $14.00
ISBN 0-688-03745-3   (library binding)   13.93
☆ Single concept, PS–1

The "one more" concept is emphasized in this counting book as a baby owl learns to count to ten. Each time the owl counts the last duck, another duck appears; so the counting becomes "Two ducks, another duck ... / Three ducks, another duck ... / Four ducks, another duck...." Class-inclusion concepts are introduced in the book when a swan appears. The baby owl is reminded to count ducks only. Simple colorful drawings illustrate the story.

Raffi. *Five Little Ducks.* Illustrated by Jose Aruego and Ariane Dewey. New York: Crown Publishers, 1989.
See Early Number Concepts: Chants, Songs, and Rhymes

Rees, Mary. *Ten in a Bed.* Boston: Little, Brown & Co., 1988.
See Early Number Concepts: Chants, Songs, and Rhymes

Richardson, John. *Ten Bears in a Bed: A Pop-up Counting Book.* New York: Hyperion Books for Children, 1992.
See Early Number Concepts: Chants, Songs, and Rhymes

Rocklin, Joanne. *Musical Chairs and Dancing Bears.* Illustrated by Laure de Matharel. New York: Henry Holt & Co., 1993.
ISBN 0-8050-2374-7 $14.95
★★★ Single concept, PS–K (PS–1)

The game of musical chairs becomes the story line for this cleverly designed book. The book opens with ten dancing bears but only nine chairs. As the waltz ends, there are nine seated bears and one lone standing bear. One by one the bears are eliminated because they are always missing one chair. Across the top of the page are tiny black-and-white drawings of the appropriate number of bears and one fewer chair. The artwork is appealing, with the lively bears bunny hopping, square dancing, rocking, and tangoing across the floor. The story line and the illustrations clearly depict "one more than" or "one less than" as well as beginning subtraction concepts. One of the tiny border graphics depicts six bears rather than the correct number of five bears, a small error that can be readily overlooked, or turned into a challenge, in such a charming book.

Salt, Jane. *My Giant Word and Number Book.* Illustrated by Sarah Pooley. New York: Kingfisher Books, 1992.

See Early Number Concepts: Counting—Other Number Concepts

Samton, Sheila White. *Moon to Sun.* New York: Caroline House, 1991.
ISBN 1-878093-14-2 $9.95

Single concept, PS–1

Night turns to day, as Sheila White Samton's colorful picture cutouts—1 moon, 1 star, 1 cloud, 1 tree, and so on—are added one by one to each subsequent page. Angular shapes form a bold and colorful collage; the items from one through ten are clearly depicted and easy to count. The first page is dominated by 1 crescent moon. As the illustrations become more complex, the moon becomes smaller and smaller. On the final page the tenth item, the sun, appears with a small moon almost hidden in the page's corner. Opposite each illustration is an accumulating listing of the individual items that have been added to the picture in the format shown at right. Although this formalization is not incorrect, connecting addition of ones to counting appears forced and may prove confusing to the young reader.

1 moon
1 star
+ 1 cloud
——
3

———. *On the River.* New York: Caroline House, 1991.
ISBN 1-878093-14-2 $9.95

Single concept, PS–1

Sheila White Samton's characteristic style of colorful picture cutouts illustrates yet another add-on book to ten. Her pictures build on the blank page as 1 river, 1 cloud, 1 fish, 1 boat, ..., 1 girl, 1 island appear until the surprise when all 10 objects compose the necessary ingredients for an island picnic. The increasing complexity of the scene lends itself to counting on by one more instead of counting all. As in *Moon to Sun,* a cumulative vertical list of items is written as an addition problem in a traditional format.

Sendak, Maurice. *One Was Johnny.* New York: HarperCollins Children's Books, 1962. Reissued 1991.

ISBN 0-06-025540-4   (library binding) $13.89
ISBN 0-06-443251-3   (paper)   3.95

★★ Single concept, PS–3 (PS–1)

Simple line drawings with limited color add humor to the story of Johnny, "who lived by himself." But he is soon joined by eight others, including a cat, a dog, a tiger, a monkey, and a robber. As the room becomes increasingly crowded, Johnny's reactions clearly show his dismay. To solve his problem he counts backward, and the animals retreat one by one. The story is told in rhyme: "3 was a cat who chased the rat, / 4 was a dog who came in and sat." The story line and the corresponding illustrations should appeal to the young reader.

Sheppard, Jeff. *The Right Number of Elephants.* Illustrated by Felicia Bond. New York: Harper Trophy, 1990.

ISBN 0-06-025615-X $14.95
ISBN 0-06-025616-8   (library binding)   14.89
ISBN 0-06-443299-8   (paper)   4.95

★★ Single concept, PS–3 (PS–1)

How many is the right number of elephants? Well, apparently it depends on the situation. "If you suddenly need to pull a train / out of a tunnel / … then the right number of elephants is … 10"! Subsequent situations described by a young boy call for gradually decreasing numbers of elephants until "But when you need / a very special friend / for a very special moment," the right number of elephants is 1. From 10 to 1, the predictable refrain is supported by the appropriate numeral and set of elephants. The illustrations are worth studying; their action detail and humor embellish this whimsical story. One charming scene is introduced by "And when you go to the beach / … and you simply must have shade, / then the right number of elephants is…." In the upper right corner is a large numeral 8; eight elephants, attired in sunglasses and sipping sodas, stand in line with their bulk shading the young boy and his friends playing on the sandy beach.

Szekeres, Cyndy. *Cyndy Szekeres' Counting Book One to Ten.* Racine, Wis.: Western Publishing Co., 1984.

ISBN 0-307-12141-0 $3.50
ISBN 0-307-12140-0

★ Single concept, K

Vivacious, charming little mice cheerfully represent the numbers from one through ten in this slender, yet sturdy, cardboard counting book. A simple text introduces the appropriate numeral and number name on each double page. The detailed pastel illustrations of the environment and the antics of the mice should delight the young reader.

Thaler, Mike. *Seven Little Hippos.* Illustrated by Jerry Smath. New York: Simon & Schuster Books for Young Readers, 1991.

See Early Number Concepts: Chants, Songs, and Rhymes

Time-Life for Children Staff. *How Many Hippos? A Mix-&-Match Counting Book.* Illustrated by Robert Cremins. Alexandria, Va.: Time-Life, 1990.

See Early Number Concepts: Counting—Other Number Concepts

Wakefield, Ali. *Those Calculating Crows!* Illustrated by Christy Hale. New York: Simon & Schuster Books for Young Readers, 1996.

See Early Number Concepts: Counting—Other Number Concepts.

Walsh, Ellen Stoll. *Mouse Count.* San Diego, Calif.: Harcourt Brace & Co., 1991.

ISBN 0-15-256023-8 $12.00

★★ Single concept, PS–1

Opportunities to "count on" are provided in this clever story involving ten mice and a snake. Searching for dinner, the snake encounters three little sleeping mice. As he drops them one by one into a glass jar, he counts, "Mouse Count! One ... / two ... / three." The snake finds four more and counts on, "four ... / five ... / six ... / seven." After all ten mice have been captured, they outwit the snake and escape from the jar; this action promotes counting backward from ten. The simple text is complemented by collage illustrations in greens, browns, grays, and blues.

Weston, Martha. *Bea's 4 Bears.* New York: Clarion Books, 1992.

ISBN 0-395-57791-8 $9.95

Single concept, PS–K

Beginning addition and subtraction concepts are embedded in the story of Bea and her four teddy bears. Bea fills her red wagon with her bears but must leave one with her dog, Bingo, because the wagon is too full. On a picnic, Silly Bear falls in the peanut butter; Bea cleans him up and leaves him on the line to dry. Only two bears are left as Bea sits down to play with her friend Kay. Events continue to gradually unfold until Bea is left with zero bears. As she searches and then pauses to think about the situation, Bea retraces her steps to recover all four bears. Framed in a pastel square border, the pleasant illustrations complement the simple text.

Wing, Natasha. *Hippity Hop, Frog on Top.* Illustrated by DeLoss McGraw. New York: Simon & Schuster Books for Young Readers, 1994.

ISBN 0-671-87045-9 $15.00

☆ Single concept, K–1

Curiosity is almost the undoing of nine nosy frogs who one by one happily hop their way on top of each other until there are "9 frogs, nine frogs tall could still not see over the wall." The addition of a tiny tree frog is just enough to quench their curiosity. The surprise ending causes havoc as the frogs in the tower jump for their lives! The wall is silhouetted against a setting sun, and the text concludes with "Hippity hop, frogs *kerplop*. Now there are no frogs against the wall." Lively prose and repetitive verse complement the artwork and incorporate both numerals and word names for the numbers from one through ten. Illustrations in gouache and colored pencil are dramatic, with their flamboyant richness of color and random-appearing spots of spattered paint, but this busy design may be distracting to young children. Other books by this illustrator have been described as "feasts of luscious color." The tall, narrow dimensions of the book add to the theme of building a tower to see over a wall.

Wise, William. *Ten Sly Piranhas: A Counting Story in Reverse (A Tale of Wickedness—and Worse!).* Illustrated by Victoria Chess. New York: Dial Books for Young Readers, 1993.

ISBN 0-8037-1200-6 $14.99
ISBN 0-8037-1201-4   (library binding)   13.89

★★ Single concept, PS–3

"And with a gulp, and a gurgle" ten fluorescent pink piranhas disappear one by one in this tale involving counting backward. One piranha stops to eat beetles on the river floor; another grows sleepy in the sunlight. Each situation results in one fewer piranha to be described in the repetitive verse. The well-written prose skillfully reflects the wicked antics of the sly piranhas. The diminishing piranhas can be readily counted as they swim in the river. The Amazonian culture is reflected by the colorful and intriguing birds, slithering and crawling animals, and exotic plants that populate the attractive illustrations. Facts about piranhas are woven into the text and the author notes at the book's end. The gradually decreasing set is a good introduction to counting backward and beginning subtraction concepts.

Counting—Other Number Concepts

Books in this final counting category are placed here because to be used successfully, they require of the reader an understanding beyond simple counting and a higher level of confidence in knowledge of enumeration. Both Audry-Iljic and Courtin's *My First Numbers* and Salt's *My Giant Word and Number Book* present counting and many other topics for the younger reader to explore. Children and adults will ponder over Geisert's *Pigs from 1 to 10,* a picture puzzle book. Other types of puzzles are found in Bourke's *Eye Count: A Book of Counting Puzzles* in which the primary theme is word play.

Other more sophisticated opportunities include counting partials within a set, such as Testa's *If You Take a Pencil.* Readers are challenged to determine how many by counting, comparing sets, and exploring one-to-one correspondence in Miller's *Farm Counting Book.* Activities in Gigante's *How Many Snails?* and Grande Tabor's *Cincuenta en la Cebra: Contando con los Animales / Fifty on the Zebra: Counting with the Animals* involve the challenges of classifying and counting part of a set. Careful discrimination is also needed in Owens's *Counting Cranes.*

Nonsequential counting is encountered in books by Allington, Carle, and Wylie and Wylie. Skip counting is highlighted in Pinczes's *Arctic Fives Arrive* and Hamm's *How Many Feet in the Bed?* Odd and even numbers are featured in Ryan and Pallotta's *The Crayon Counting Book.* Patterns in some books lead to preliminary ideas in number theory. Primes and composites are implied in Wahl's photographs of flowers in *I Can Count the Petals of a Flower.* All these books have one common idea—using counting as a point of departure to explore more sophisticated mathematical ideas.

Allington, Richard L. *Numbers.* Illustrated by Tom Garcia. Chatham, N.J.: Raintree Steck-Vaughn Publishers, 1985.

ISBN 0-8172-1278-7 $9.95
ISBN 0-8114-8239-1   (paper)   3.95

★★ Single concept, K–2

After an introductory two-page spread giving the numerals 1 through 10 and the corresponding sets, the reader becomes involved in counting colorful sets of animals to determine how many there are. Counting is necessary because the book is designed so that the numbers are not presented in sequence, and the numeral answer to each question is found by turning the page. In another activity, the reader is challenged to find numerals that are cleverly integrated into an illustration of a house and its surrounding yard. In a matching activity, recognition of the numerals and their meaning is reinforced by having the reader match the numerals 1 through 10 with new sets of animals. The book's final page suggests two follow-up activities for making a number book or building a number collection.

Audry-Iljic, Françoise, and **Thierry Courtin.** *My First Numbers.* Adapted by Judith Herbst. Hauppauge, N.Y.: Barron's Educational Series, 1994.

ISBN 0-8120-6314-7 $12.95

★★ Multiconcept, PS (PS–1)

Translated from a French edition, this book explores partitions of numbers as well as counting. Illustrations are simply done with minimal detail in strong colors against a solid-colored background. The well-written text is in poetry form with irregular rhyming schemes. Bold print highlights the featured number name. In the first section of the book a little boy introduces the numbers from one through ten. A large numeral dominates the left-hand page, with verse highlighting its shape. The opposite page effectively depicts the child with an appropriate number of counters; for example, six baby chicks are shown as "**Five** on this side, **one** on the other, / that makes **six** chicks looking for their mother." Additionally, on each sequential layout a large spherical counter is threaded on a rod to form a growing column. The first five counters are red; the next are all green so that the reader can clearly count five and one more, five and two more, and so on. The book's second section explores number in different ways. Examples include ten bowling pins knocked down to indicate that zero pins are standing and an illustration of two adult teddy bears and a small teddy bear that mirrors the domino pattern shown on the corner of the page. Six is shown as three sets of two. A pile of seven presents is rearranged into a stack; the patterns on the dominoes show 4 and 3 then 5 and 2. The foreword includes a note to the adult reader that describes children's gradual acquisition of number concepts.

Bourke, Linda. *Eye Count: A Book of Counting Puzzles.* San Francisco, Calif.: Chronicle Books, 1995.

ISBN 0-8118-0732-0 $13.95

★★ Single concept, all ages

Eye Count is a mystery counting book; the reader must discern what is to be counted. Each intriguing illustration contains many different objects, but the key is determining what they have in common. For the number four, the common element is homonyns, with a jack of hearts playing card, a jack-in-the-box, a ball and jack, and a telephone jack clearly featured. Each illustration contains a hint for the next puzzle; for the page depicting four, a telephone cord dangles near the page's edge. For the number five, homophones for *chord* are the focus; represented in the illustration are a musical chord, a cord of wood, an apple core, a telephone cord, and a cord for window blinds. The attractive illustrations were done with Prismacolor pencil on black paper. Across the top of each double page are the numeral and the number word name. Counting is merely the vehicle used to focus on word use. Solutions for the ten puzzles are included.

Carle, Eric. *My Very First Book of Numbers.* Reprint. New York: Crowell Junior Books, 1985.

ISBN 0-694-00012-4 $4.95

☆ Single concept, PS–K

Carle develops this book in the same format as *My Very First Book of Shapes.* The cardboard pages are cut horizontally with the numerals from 1 through 10 and the appropriate number of black squares on the top half of the page. Vibrant pictures of fruit in sets of one through ten are pictured on the bottom half of the pages. The numeral pages are ordered sequentially, whereas the pictures of fruit are ordered randomly. The child can turn the separate halves of the pages to match the pictures with the appropriate numeral and number of counters.

Geisert, Arthur. *Pigs from 1 to 10.* Boston: Houghton Mifflin, 1992.

ISBN 0-395-58519-8 $14.95

★★ Single concept, PS–3

Geisert's intricate etchings on a pale cream background reflect the fantasy and mystery of the ten pigs' quest to find stone configurations in a lost location. Each scene shows them hard at work as they struggle with a cannon, build a bridge, and drill through rocks until finally they discover the place described by their mother. Finding and counting each of the ten pigs in each illustration is no casual task. A greater challenge is to find each numeral from 0 through 9 that is are carefully designed into each illustration. The steam from a machine may form a 6; the design of bricks, a 1; or the coil of a rope, an 8. The designs are subtle; a key to the illustrations is included. Finally the mystery is solved when the pigs discover that the strange rock formations are the numerals from 0 through 9. After tracing each numeral, the pigs carry the wooden copies across the bridge to their home. Because of its sophisticated design, this elegant book is foremost a picture puzzle book and secondarily a counting book.

Giganti, Paul, Jr. *How Many Snails? A Counting Book.* Illustrated by Donald Crews. New York: Greenwillow Books, 1988.

ISBN 0-688-06369-1		$15.00
ISBN 0-688-06370-5	(library binding)	14.93
ISBN 0-688-13639-7	(paper, Mulberry)	4.95

☆ Single concept, PS–1

Gouache paints were used to create the simple, colorful illustrations of objects for the reader to classify and count. On each double page, several questions are posed about the large set depicted there. On a page covered with a school of fish, the questions challenge the reader to find how many fish there are, how many fish are red, and how many red fish have their mouths open. The sets, which vary in size up to forty-three, provide discrimination opportunities in color, size, number, and design.

Grande Tabor, Nancy Maria. *Cincuenta en la cebra: Contando con los animales / Fifty on the Zebra: Counting with the Animals.* Watertown, Mass.: Charlesbridge Publishing, 1994.

ISBN 0-88106-858-6		$15.88
ISBN 0-88106-856-X	(paper)	6.95

☆ Multiconcept, K–2

Spanish and English appear side by side in two columns as number values are explored from 0 through 20, then by tens to 100. A statement in bold print focuses the reader on the number to be explored and counted: "Zero snakes begin the race." Beneath each statement, four or five questions are posed. Some questions relate to numerical understanding; others are open-ended, for example, "How many feet does the snake have? How many leaves are on the winter tree? Who do you think will win the race? Which season of the year do you like best? What will the prize be at the end of the year?" In the accompanying illustration, four trees depicting the four seasons grow along a winding path on which a rabbit and a tortoise are racing; a snake is posed at the path's beginning. This format is followed throughout as increasingly large numbers of animals are introduced. Numerals are rarely seen. Questions on several pages encourage readers to classify, combine, and compare members of different sets. Often the reader is asked questions like "How many turtles are green?" or "Are there more swans or more llamas?" Tabor's printed collages are done in brilliant color; for both design purposes and ease of counting, symmetry and groupings are effectively used in several illustrations.

Grossman, Bill. *My Little Sister Ate One Hare.* Illustrated by Kevin Hawkes. New York: Crown Publishers, 1996.

See Early Number Concepts: Counting—One to Ten

Hamm, Diane Johnston. *How Many Feet in the Bed?* Illustrated by Katie Salley Palmer. New York: Simon & Schuster Books for Young Readers, 1994.

ISBN 0-671-89903-1 $4.95

★★ Single concept, (PS–1)

   Pleasant early morning scenes from a family's life are captured in Palmer's charming illustrations. One child's question, "How many feet in the bed?" becomes the theme. When her father replies that there are two, she quickly jumps in to make four. When joined by her little brother, they count by ones to six. Counting by ones, and later by twos, continues until ten feet are in bed. Subtracting by twos starts as one by one each family member gets out of bed. This delightful book could be used as a springboard for counting by twos or tens (toes)—forward or backward—or for adding and subtracting.

Hoban, Tana. *Twenty-six Letters and Ninety-nine Cents.* New York: Greenwillow Books, 1987.

See Measurement: Money

McGrath, Barbara Barbieri. *The "M&M's" Brand Chocolate Candies Counting Book.* Watertown, Mass.: Charlesbridge Publishing, 1994.

ISBN 0-88106-854-3   (hardcover) $14.95
ISBN 0-88106-855-1   (library binding)   15.88
ISBN 0-88106-853-5   (paper)   6.95

☆ Multiconcept, PS (PS–2)

   Blue, green, orange, yellow, red, and brown M&M's candies are used to illustrate each of the numbers from one through six. Numbers from seven through twelve are represented by adding new colors one by one to a row of six brown candies; each new addition creates a new pattern. Color photographs on a plain white background and simple rhyming text introduce each new idea. Word names are written in large text, and numerals are written in crayon. Rearrangements of the twelve candies effectively illustrate other names for twelve, such as six sets of two, three sets of four, four sets of three, and six sets of two. Bold type announces the number of sets and the related addition problem is written in crayon. The twelve candies are rearranged to form the geometric shapes of square, circle, and triangle. Subtraction is illustrated by having one blue candy disappear, then all seven brown candies, until finally all the M&M's are gone. The related subtraction problems are recorded in crayon.

Children will surely enjoy the subtraction section, which starts with the following invitation: "Now comes the part that will be the most fun. We'll start to subtract—so eat the blue one." A review page reinforces the concepts of color, shape, number names, numerals, and equal-sized sets to represent twelve.

Miller, Jane. *Farm Counting Book.* New York: Simon & Schuster Books for Young Readers, 1992.
ISBN 0-671-66552-9 $5.95
☆ Single concept, PS–3

Familiar farm animals and objects appear in colorful photographs. The numerals and number names for one through ten accompany such farm scenes as five cows in a pasture and eight swans in a pond. Additional pages involve the reader in such activities as counting different sets, comparing sets, and exploring one-to-one correspondence by deciding if there are enough bowls for the cats. One flaw is the picture representing seven, which shows seven ducks and a partial picture of two more ducks.

Moncure, Jane Belk. *The Magic Moon Machine.* Illustrated by Linda Hohag. Chicago: Children's World, 1987.
ISBN 0-89565-920-4 $21.36
ISBN 0-89565-041-5 21.36
☆ Single concept, PS–2

Kim wants to join the astronaut in the Magic Moon Machine but keeps remembering one more thing he needs to take with him on his trip. It is not long before the machine is so heavy and overcrowded with Kim's pets, toys, food items, and furnishings for a house that it cannot leave the ground. The classification of items by type is clearly defined. Unlike in most counting books, the numbers are not introduced in sequence. Kim takes his pony, two cats, four dogs, a bowl with two goldfish, and eight jumping frogs. The review pages at the end of the book help the reader recall the items by numeral and matching number set. The closing question, "What would you take to the moon with you?" could motivate children to choose different categories and list the items in each one.

My First Number Book. New York: Dorling Kindersley, 1992.
See Number—Extensions and Connections: Number Concepts

Owens, Mary Beth. *Counting Cranes.* Boston: Little, Brown & Co., 1993.
ISBN 0-316-67719-1 $14.95
★★ Single concept, PS–3 (all ages)

Watercolors and acrylics in soft natural tones and beautiful blank verse combine to capture the life and migration of whooping cranes. The cranes are introduced from one through fifteen; only number word names are used. Often the reader must carefully discriminate among the cranes and other waterfowl and birds when counting. Counting becomes secondary in this book designed to honor the North American whooping crane. The cranes' habitats and experiences are depicted as they migrate from Canada to the southern United States. Maps of the cranes' journey are included in the endpapers, and a note from the author-artist includes information about this endangered species.

Pinczes, Elinor J. *Arctic Fives Arrive.* Illustrated by Holly Berry. Boston: Houghton Mifflin Co., 1996.

ISBN 0-395-73577-7 $14.95

☆ Single concept, PS–3 (PS–2)

"One late Arctic day, not too far from the bay, / five snowy owls gracefully flew. / They spotted a mound, the tallest around, / where small birds could have a nice view." Rhyme introduces each new group of five animals as they gather on top of a tall piece of ice drifting in the chilly northern waters. Each new group—five snowy owls, five polar bears, five sly ermines, ..., five musk oxen—generates a count by fives: five, ten, fifteen, ..., thirty. Each figure, outlined in black, is drawn with minimal detail; huddled together on the narrow mass, individual members of the accumulating group can be difficult to identify, but the intent is counting by fives, not by ones. Dominated by Arctic white and pastel shades of blues, purples, and pinks, the illustrations reflect the approach of the day's end. The story concludes as the reason for the animals' gathering is unveiled—the appearance of the northern lights—and the animals quietly disperse: thirty, twenty-five, twenty, fifteen, ten, and five. Counting sequences are also reflected in the white numerals printed on purple borders that frame some of the illustrations. Mathematically, *Arctic Fives Arrive* is not as rich as Pinczes's previous books, but it could be used to generate a discussion of counting by fives as well as an exploration of counting by other multiples.

Pluckrose, Henry. *Counting.* Illustrated by Chris Fairclough. Chicago: Children's Press, 1995.

See Series and Other Resources: Math Counts

Potter, Beatrix. *Peter Rabbit's 1 2 3.* New York: Frederick Warne, 1987.

ISBN 0-7232-3424-8 $6.95

☆ Single concept, PS–3

Peter Rabbit and his friends provide an opportunity for the reader to count from one through twelve. Each number is introduced by having the reader answer a question about the activities of Peter and his friends. On the same page is the numeral and the appropriate number of animals. On the facing page is the number name and a detailed illustration of these animals as they appear in the original illustrations by Beatrix Potter. An appropriate quote from the original story accompanies the illustrations. This is not an introductory counting book because the correct number of characters is often difficult to find in the illustrations from the original text. For this reason, only the competent counter should be given the challenge of this book. Additionally, the end papers showing the numerals 1 through 12 could be confusing because the sketches of the animals are solely decorative rather than related to counting. Questions at the end of the book offer further opportunities for counting. A Spanish edition of *Peter Rabbit's 1 2 3* is also available.

Ryan, Pam Muñoz, and **Jerry Pallotta.** *The Crayon Counting Book.* Illustrated by Frank Pallotta, Jr. New York: Charlesbridge, 1996.

ISBN 0-88106-954-X	$14.95
ISBN 0-88106-955-8 (library reinforced)	15.88
ISBN 0-88106-953-1 (paper)	6.95

★★ Single concept, (PS–2)

The even numbers from zero through twenty-four and the odd numbers from 1 through 23 are clearly shown as two more crayons (sometomes used or broken!) are added to represent each new set. Often the crayons are arranged in a pattern: eight is shown as two groups of four; eleven, as three rows of two plus two more; nineteen, as standing columns of ten and nine. Symmetry is found in some of the patterns. Each new number is introduced by a rhyming verse, for example, "Welcome the buddies peach and gray, / and discover a dozen in the display." Each number word name and numeral is printed across the top of each page in an outline font; for each new layout, the interiors are colored to match the two new crayons. Neither the word name *zero* nor the symbol 0 has been colored on the opening page where zero is featured as an empty crayon box. After the introduction of the evens, a summary page shows twelve pairs of crayons with each word and numeral for the counting sequence. A similar summary follows the introduction of the odd numbers along with a reminder that the reader has counted in twos in two different ways. Readers should find this well-designed book most appealing.

Salt, Jane. *My Giant Word and Number Book.* Illustrated by Sarah Pooley. New York: Kingfisher Books, 1992.

ISBN 1-85697-861-3 $9.95

☆ Multiconcept, K–2

My Giant Word and Number Book covers a lot of territory—in vocabulary of everyday words and in mathematics concepts. Because the success of many activities depends on adult interpretation, parentd and child should work together and heed the author's own words: "Above all, this is a book to enjoy together." Every page, including the end pages, is full of busy children, scattered text, balloons indicating direct speech, and numerals. The details in the illustrations demand a careful look to catch every nuance. Each page has a theme that relates to a young child's own interests and experiences. Each theme incorporates mathematics concepts in realistic settings: one-to-one correspondence in counting, numerals as identifiers, number and time, more and less, measurement, bigger and smaller, shapes, and number sentences. Zero is effectively introduced in a five-page sequence with the numbers from zero to ten. "Zero, one, two, three"—across the top of the first page—are the counting names for the introduced numbers. Groups of toys are to be counted for each number. Across the bottom of the page, the same groups of toys are shown arranged in four carts, with each cart labeled with a numeral from 0 to 3. To accommodate the increasing numbers, new and larger carts are added to the train as it progresses across each page. Simple addition and subtraction sentences involving one more or one less are effectively introduced through well-designed illustrations and text. On a few occasions the busyness of the illustrations becomes confusing. Children representing various ethnic backgrounds are shown in the cartoonlike illustrations; a table of contents is included.

Testa, Fulvio. *If You Take a Pencil.* New York: Dial Books for Young Readers, 1993.
ISBN 0-8037-4023-9 $5.95
ISBN 0-8037-0165-9   (paper, 1985)   4.95
☆ Single concept, PS–2

Although this book is designated a counting book, its value lies in the colorful illustrations produced from ink-and-dye paintings. It is a book for a more experienced reader, since the counting often involves partially hidden objects. The illustrations and the story line are fanciful because they come from the ideas a reader can explore through imagination and drawing. The closing pages are an invitation by the author to explore through the reader's own drawing.

Time-Life for Children Staff. *How Many Hippos? A Mix-&-Match Counting Book.* Illustrated by Robert Cremins. Alexandria, Va.: Time-Life, 1990.
ISBN 0-8094-9258-X
ISBN 0-8094-9259-8   (library binding)
★★ Multiconcept, PS–2

More than a traditional counting book, this book, through its unique format, encourages creative tasks for counting and addition. Counting from 1 through 10 is the purpose of the first section. Each set of objects is depicted against a white background and, with the appropriate numeral, is introduced on the top half of horizontally split pages. Readers are encouraged to create a counting caravan by matching the bottom halves of the pages with the top halves. Choices include 2 fancy flamingos balancing on a panda's finger, teetering on the seal's nose, or joining a bear's juggling act. The second section of the book can be used to compare, add, or subtract two sets. Both the top and the bottom halves of the book depict the sets from one through ten. There are often numerous sets to count on each half page, but a set of red birds is always present. On the edge of each page is a vertical row of ten squares; the number of red squares matches the number of red birds. Questions on these pages soon involve the reader in comparing the sets or answering questions with addition or subtraction. By flipping the top and bottom pages back and forth, the reader can solve new problems. On a final double page, numbers from 11 through 20 are represented by vertical rows of red squares, arranged in sets of five; the reader is encouraged to count by fives to verify how many. Included in both sections is a note to parents on how to use the book. The cartoonlike illustrations are brightly colored and appealing. This book is part of a series, Time-Life Early Learning Program.

Wahl, John, and **Stacey Wahl.** *I Can Count the Petals of a Flower.* Reston, Va.: National Council of Teachers of Mathematics, 1985.
ISBN 0-87353-224-4 $8.00
★★ Multiconcept, K–4

Colored photographs of flowers clearly illustrate the concepts of one through sixteen in this unique book. The authors have combined a variety of techniques to reinforce number concepts and patterns. Each concept of one through ten is depicted by a photograph of a flower, the appropriate number symbol to identify the number of petals on the flower, and beneath it a simplified abstract sketch of the flower that emphasizes the petals only. Subsequent pages count from one through sixteen and feature single flowers or combinations of flowers whose petals represent the number indicated. An interesting technique deals with prime numbers. If a number is prime, such as seven, one flower is shown; but for eight, which is not prime, there are pictures of an eight-petaled flower, four two-petaled flowers, and two four-petaled flowers. Toward the end of the book, the reader is shown several flower photographs and asked to solve simple counting problems. The reader may be tricked on the last page where she or he is asked, "How many petals do I have?" The five photographs depict plants whose flowers do not have the same number of petals even though they are on the same plant.

Wakefield, Ali. *Those Calculating Crows!* Illustrated by Christy Hale. New York: Simon & Schuster Books for Young Readers, 1996.
ISBN 0-689-80483-0 $16.00
☆ Single concept, K–3

Can crows count? Counting is viewed as a human activity, but

this unusual counting book reports an experiment that hunters
conducted many years ago. The story unfolds as a farmer
contemplates how to scare the crows away from his newly planted
cornfield. He hides in the barn, but the guard crow warns the other
crows to fly away; they return when the farmer leaves. A plan is
devised; two people will enter the barn and only one person will
leave. It appears that the crows know how to count, since they stay
away. The experiment continues as three people enter the barn and
two leave; four people enter and three leave, and so on. Each time, it
seems that the crows can count, since they stay away until the extra
person leaves the barn. Various-sized illustrations are effectively
interspersed with the text to tell both the farmer's story and the
crows' story; the attractive, well-designed illustrations enhance the
humor of the story. Opportunities to explore number include
counting, adding, and subtracting as well as developing number
sense. Numerals and numbers from one through eight appear with
an appropriate number of crows to count. Fanciful illustrations of
the crows' activities depict them checking computer monitors for
grain prices and reading the Crow's Feet Journal! Repetitive rhyme
adds to the humor on these pages.

Walton, Rick. *How Many, How Many, How Many.* Illustrated by Cynthia
Jabar. Cambridge, Mass.: Candlewick Press, 1993.

ISBN 1-56402-062-2 $14.95
ISBN 1-56402-656-6   (paper)   5.99

★★ Single concept, PS–1

"Are you ready? Tell me yes. / How many HOW MANYs can you
guess?" Twelve children, each holding a numeral from one through
twelve, stand behind the starting line ready to go! This first double-
page illustration sets the tone for a thought-provoking and
informative race to answer each "how many?" question. Numbers are
sequentially presented with a large colored numeral and the
corresponding word name prominently displayed on each two-page
layout. "Curds and whey" are the two objects to be counted to answer
"How many things did Muffet eat?" The seven colors in a rainbow,
Santa's eight reindeer, and the nine planets orbiting the sun are
named and clearly illustrated for ease in counting. One thought-
provoking example for the young child involves the ten numbers on a
telephone—the digits from zero through nine. The lively scratchboard
illustrations are charming and humorous. Children representing
different races are depicted; both girls and boys are shown cheerfully
playing on the soccer team in one lively scene. The last double-page
spread depicts the finish line of the race where competitors are in
various stages of delight, exhaustion, or both. But wait! Don't rest
yet, there's a final challenge: "And were you looking carefully? /
Somewhere did you also see...." On the last single page is a
chalkboard that calls for an additional search: one Tom Thumb, two
spiders,..., twelve hours. Here we go again!

Wylie, Joanne, and **David Wylie.** *Cuantos monstruos? Un cuento de números.* Spanish ed. Chicago: Children's Press, 1988.

ISBN 0-516-34494-3   (library binding) $11.93
ISBN 0-516-54494-2   (paper)   3.95

★★ Single concept, PS–2

Kindergarten children will love to count the zany monsters in this inventive book in the Many Monsters series. However, the young readers must be beyond the rote-counting stage to fully enjoy the challenge of this book. They are asked to discriminate what is to be included before they count what appears on each page. For example, readers are asked to count kneeling monsters only, but some monsters who are not kneeling have bent knees, and youngsters may have difficulty seeing the difference. Answers in numeral form are presented by a tiny monster on each page. The sets to be counted in this unique book are not sequenced, and the questions to be explored are posed in amusing rhymes. The comical characters are depicted in vibrant color to provide a visually pleasing and challenging experience. Only the Spanish edition of *How Many Monsters? Learning about Counting* appears to be available.

Number—
Extensions and Connections

W HEN a class of college students were asked to explain what 5! meant, they told how to compute it; that is, 5! is $5 \times 4 \times 3 \times 2 \times 1$. They could not describe a representation of the concept or give any explanation that indicated a working understanding of the concept other than a vague reference to "Maybe it is used in probability." *Anno's Mysterious Multiplying Jar* was then read to the class. The discussion of the book by these preservice teachers included the quality of the representation of the concept and how they could use the book with their future students. Several students commented that they wished such experiences had been part of their elementary school background. They also expressed a need to have such books shared with them—both for their concept development and for their awareness of what is available for elementary school children.

In Anno's books, as well as in many others like *One Hundred Hungry Ants* or *Math Curse,* sophisticated mathematics is often couched in the context of a problem-solving situation. These books provide excellent opportunities for children to explore mathematics concepts in an intuitive, informal setting. These experiences should provide motivation and a stepping-stone for children to continue exploring the world of mathematics. The theme of the books in this section is exploring number. Some books are appropriate for very young children who are still developing early number ideas, but most books introduce young readers to more sophisticated mathematics concepts. This category has six subcategories. "Number Concepts" contains an extensive group of books that explore number sense, number use, estimation, and fractions. Cultural and historical connections to numeration systems are the basis of another category, "Number Systems." A third category, "Operations," includes many excellent explorations of operations: concepts, properties, story problems, and number and operation sense. Connections to the real world through data collection and chance are found in "Statistics and Probability." A fifth category, "Number Theory, Patterns, and Relationships," includes books that explore the patterns and mysteries of number theory and such relationships as proportionality. In the books in "Reasoning— Classification, Logic, and Strategy Games," logical reasoning and classification are explored directly; those concepts are also found in strategy games. Additional classroom vignettes illustrate how some of these books have been used successfully with children.

Number Concepts

The first graders expressed delight and amazement as their teacher, Sue Sherwood, read Schwartz's *How Much Is a Million?* As they talked about the size of numbers, one child asked about the largest number. Two children quickly interjected that there isn't a largest number because there is always a number that is one more. As the children interacted, one could see that at least three first graders were starting to comprehend the concept of infinity.

Number sense and comparisons for large numbers in the millions, billions, and trillions are illustrated by Schwartz in *How Much Is a Million?*, an excellent example of a book that can be successfully used and enjoyed by children of many ages. Number sense, including an appreciation for the relative size of numbers, is encountered in many books in this category; refer to the category "Counting—Other Number Concepts" for additional simpler books that develop cardinal numbers.

Other aspects of number concepts are also included in this section. Names for, and concepts of, special groups are illustrated in Harshman's *Only One* and McMillan's *One, Two, One Pair!* Merriam's *Twelve Ways to Get Eleven* is an excellent example of partitioning numbers and readily leads children to explore and write about other numbers. Number concepts are extended to introduce fractions through representations by Adler in *Fraction Fun,* Mathews in *Gator Pie,* and Leedy in *Fraction Action.* Number *use* with respect to whole numbers and common fractions also appears in this exploration of number concepts. Some books on whole numbers focus on how numbers are used for ordering and for naming. Ordinal concepts are illustrated in *My First Number Book;* nominal concepts, or naming, such as street addresses, are explored in Moore's *Six-Dinner Sid* and Pluckrose's *Numbers.* Other connections to number uses in real life are illustrated by Aker in *What Comes in 2's, 3's, and 4's?*

The following books are out-of-print but are considered exceptional:

Dennis, Richard. *Fractions Are Parts of Things*

Froman, Robert. *Less Than Nothing Is Really Something*
Gillen, Patricia Bellan. *My Signing Book of Numbers*
Hellen, Nancy. *The Bus Stop*
Maestro, Betsy. *Harriet Goes to the Circus: A Number Concept Book*
Sitomer, Mindel, and **Harry Sitomer.** *Zero Is Not Nothing*

Copies may be available in your library. See the first edition of *The Wonderful World of Mathematics* for annotations.

Adler, David A. *Fraction Fun.* Illustrated by Nancy Tobin. New York: Holiday House, 1996.
ISBN 0-8234-1259-8 $15.95
★★★ Single concept, PS–3 (1–4)

Connections to our world are central in this thoughtful introduction to fractions. Concepts of equal-sized parts, word and numeral names, and the meaning of the numerator and denominator are discussed with examples of pizza slices. Paper plates, crayons, and rulers are needed for constructing models for halves, fourths, and eighths; the tasks are designed to help children formulate generalizations after they explore whether one-half is larger than one-fourth or three-eighths is more than two-eighths or one-eighth. In another activity, children weigh objects like pennies, tissues, or envelopes and are again led to generalize: since six nickels weigh an ounce, one nickel must weigh one-sixth of an ounce. A final task uses graph paper for finding equivalent fractions. Cartoon characters and bright colors effectively illustrate each concept. *Fraction Fun* should appeal to children and adults; Adler's selection and presentation of tasks for this exploration of fraction concepts are excellent.

Aker, Suzanne. *What Comes in 2's, 3's, and 4's?* Illustrated by Bernie Karlin. New York: Simon & Schuster, 1990.
ISBN 0-671-67173-1 $13.95
ISBN 0-671-79247-4   (paper)   4.95
☆ Single concept, PS–K (PS–1)

Eyes, ears, arms, hands, legs, feet, handles on a sink, and pieces of bread for a sandwich are examples given for "What comes in 2's?" The examples throughout the book are familiar and varied. Items to show "three-ness" include traffic lights, wheels on a tricycle, meals each day, and primary colors. In addition to examples that can be observed, such as wheels on a car, the book includes examples of ideas, such as seasons of a year. Colorful illustrations complement a simple text that should motivate the reader to find other things that come in twos, threes, or fours.

Anno, Mitsumasa. *Anno's Math Games II.* New York: Philomel Books, 1989.

See Number—Extensions and Connections: Number Theory, Patterns, and Relationships

Ash, Russell. *Incredible Comparisons.* New York: Dorling Kindersley, 1996.

See Number—Extensions and Connections: Statistics and Probability

Atherlay, Sara. *Math in the Bath (and other fun places, too!).* Illustrated by Megan Halsey. New York: Simon & Schuster Books for Young Readers, 1995.

ISBN 0-689-880318-4 $12.00

☆ Multiconcept, PS–2

Mathematics is all around us in our daily lives and often goes unnoticed; making students aware of the pervasiveness of mathematics is the reason the author designed this book. From getting up in the morning to going to bed at night, the book traces the child's day. Examples of connections between mathematics and other school subjects include patterns in music, geometric shapes and color mixing in art, and flag designs in social studies. Measuring and counting are embedded in the games at recess. The illustrations provide a simple and uncluttered complement to the text. In a note to parents and teachers, the author observes that the book was the result of encouraging her students to create their own stories describing how mathematics is all around them.

Berenstain, Stanley, and **Janice Berenstain.** *Bears on Wheels.* New York: Random House, 1969.

ISBN 0-394-80967-X $6.95
ISBN 0-394-90967-4   (library binding)   9.99

★★ Single concept, PS–1

Slapstick humor and outrageous action pervade this "typically Berenstain" book, as bears on unicycles speed across each page. The colorful pictures show such unlikely sights as "two bears on one wheel," "four on one," "two on two," "one on three," and the anticipated "four on none," as the bears crash crazily to the ground. This book uses counting for its delightful fast-paced introduction to comparisons.

Bertrand, Lynne. *One Day, Two Dragons.* Illustrated by Janet Street. New York: Clarkson Potter/Publishers, 1992.

ISBN 0-517-58413-1

☆ Single concept, PS–3

Numbers appear in different contexts throughout this delightful story of one day in the life of two dragons. Setting off for the doctor's office, "One day, two dragons caught a cab to Three Bug Street." The pastel illustration shows two dragons leaving their home; the numeral 3 appears on a car license plate, a 2 can be seen on a road sign, and a newspaper advertises a 1 cent sale. Their 5 o'clock appointment is for four shots—four humorous illustrations reflect the shots for measles, mumps, scale rot, and morning breath. Some numbers are exaggerated as the two dragons deal with their fears of getting shots; the receptionist looks at least nine feet tall and the hospital hall looks fourteen miles long. Other numbers the dragons encounter name things, like room number 15. Others are countable, like the seventeen tongue depressors in the jar and the twenty lollipops. This charming story with its appealing illustrations provides an opportunity to discuss the different uses of number in our lives as well as children's medical issues.

Bulloch, Ivan. *Games.* New York: Thomson Learning, 1994.

See Series and Other Resources: Action Math

Burns, Marilyn. *Math for Smarty Pants.* Illustrated by Martha Weston. Boston: Little, Brown & Co., 1982.

See Number Extensions and Connections: Number Theory, Patterns, and Relationships

Cave, Kathryn. *Out for the Count: A Counting Adventure.* Illustrated by Chris Riddell. New York: Simon & Schuster Books for Young Readers, 1992.

See Early Number Concepts: Counting—Place Value

Clement, Rod. *Counting on Frank.* Milwaukee, Wis.: Gareth Stevens Children's Books, 1991.

ISBN 0-8368-0358-2   (library binding) $18.60

★★★ Single concept, 1–2 (1–8)

Frank (a dog) and his young owner (who wears sunglasses just like his dog's) are the stars of this outstanding book on the estimation of numbers and measurements. Each double-page spread with its hilarious illustrations shows the two companions investigating new and outlandish situations. One continuous line spiraling over the living room walls, furniture, and a sleeping dad illustrates that an average ballpoint pen will draw a line seven thousand feet long before running out of ink. Piles of "Franks" depict the estimate that twenty-four Franks should fill the bedroom. The family, including the ever present Frank, sits surrounded by green peas to illustrate the estimate, "If I had accidentally knocked fifteen peas off my plate every night for the last eight years, they would now be level with the table top. Maybe then Mom would understand that her son does not like peas." The duo also explores volume—by considering how long it would take to fill the bathroom with water—and length—by determining how tall the boy would be if he grew as fast as the gum tree in the yard. This is an excellent book for helping develop number sense, and the zany situations and illustrations should captivate young readers as well as adults.

Emberley, Ed. *Ed Emberley's Picture Pie: A Circle Drawing Book.* Boston: Little, Brown & Co., 1984.

See Geometry and Spatial Sense: Geometric Concepts and Properties

Giganti, Paul, Jr. *Notorious Numbers.* Illustrated by Aaron Grbich. San Leandro, Calif.: Watten/Poe Teaching Resource Center, 1993.

ISBN 1-56785-006-5

☆ Single concept, (K–3)

Notorious numbers are numbers that have a unique role. There are seven days in a week, so 7 is a notorious number. Similarly, five and two have special roles, since there are five fingers on each hand and two items make a pair. A specific example is given for each number from 1 through 10. The text is repetitive: "2 is a notorious number because...." Throughout, the reader is asked to think of other things that come in each specific quantity. Other featured numbers are 12—objects in a dozen, 24—hours in a day, 100—cents in a dollar. One million is introduced at the book's end with a vague connection to the stars in the sky. A lone teddy bear dressed in overalls is shown presenting each example; the illustrations are simply done, often on a plain white background.

Griffiths, Rose. *Games.* Illustrated by Peter Millard. Milwaukee, Wis.: Gareth Stevens Publishing, 1994.

See Series and Other Resources: First Step Math

———. *Numbers.* Illustrated by Peter Millard. Milwaukee, Wis.: Gareth Stevens Publishing, 1994.

See Series and Other Resources: First Step Math

Harada, Joyce. *It's the 0-1-2-3 Book.* Union City, Calif.: Heian International, 1985.

ISBN 0-89346-252-7   (paper) $7.95

★★★ Multiconcept, PS–1

Introductory and end pages present color-coded number charts from 0 through 109 surrounded by sketches of cartoonlike animals in cardinal, ordinal, and nominal situations. The text is simple; a numeral on the left-hand page with an appropriate set of counters and an action-packed and highly colorful illustration on the opposite page give opportunities to count various sets. The treatment of zero is interesting because the winter scene includes a painting of the same scene in spring; as a contrast to the spring painting, the winter scene depicts zero leaves, zero flowers, zero birds in the cage, and zero boots. The counters on the page for ten are connected and offer an appropriate lead into the subsequent pages in which counters alone represent the numbers from eleven through twenty. Each group of ten is clearly defined. Some final comments by the author include suggestions for using the book. The illustrations convey a warmth and charm that set this book apart.

Harshman, Marc. *Only One.* Illustrated by Barbara Garrison. New York: Cobblehill Books, 1993.

ISBN 0-525-65116-0 $13.99

★★ Single concept, PS–3

"There may be a million stars, / But there is only one sky.... There may be 11 cows, / But there is only one herd." Harshman's predictable prose introduces objects that compose a collection. Some collections have established meanings, such as the 12 in a dozen, the 3 in a trio, or the 10 cents in a dime. Others, such as 7 peas in a pod, 6 jewels in a necklace, or 100 patches on a quilt, offer opportunities to count, whereas the estimates of 500 seeds in a pumpkin and 50 000 bees in a hive are used to introduce large-number concepts. Each statement is printed on a white background. A narrow brown border frames the left-hand pages. The opposite pages show a country fair scene representing each number relationship. Soft brown tones dominate the pleasing illustrations. A minimal use of other muted autumn colors adds to the appeal of the scenes, which were created through a combination of collage and watercolor washes. Garrison's artwork and Harshman's prose provide an excellent opportunity to launch discussions about groups of objects and number sense.

Leedy, Loreen. *Fraction Action.* New York: Holiday House, 1994.

ISBN 0-8234-1109-5 $15.95
ISBN 0-8234-1244-X   (paper)   6.95

★★ Single concept, PS–3

Learning can be fun! This colorful book of cartoon characters really develops number concepts as readers are introduced to Miss Prime and her methods to teach and involve her students in learning about and using fractions. The book has five parts: "Fraction Action," where the class learns about halves, thirds, and fourths; "Get Ready, Get Set," where the class divides a set of marbles into fifths; "A Fair Share," where Sadie has to figure out how to divide the food into halves, fourths, and fifths; "Lemonade for Sale," where Tally subtracts fractions to sell his lemonade; and "Teacher's Test," where the class tests Miss Prime on ordering and equivalent fractions. The mathematical concepts are clearly illustrated through diagrams, cartoon action, and applications to real-life situations. Region models, such as circles, rectangles, and parallelograms carefully divided into the appropriate number of parts, introduce fraction concepts and both the word and numerical names. Conversations between the teacher and her class (an interesting assortment of animals) have a problem-solving focus as they discuss their thinking and solutions. Answers to some of the problems that Miss Prime poses are found on the chalkboard featured at the back of the book.

Mathews, Louise. *Gator Pie.* Illustrated by Jeni Bassett. Denver, Col.: Sundance, 1995.

ISBN 0-7608-0005-7   (paper) $4.95

★★★ Single concept, 1–4

Alvin, Alice, and other alligators inhabit this delightful book on beginning fraction concepts. Alvin and Alice decide to share a pie they find in the woods. But before they can cut it, other alligators arrive; so they need to consider cutting the pie in thirds, fourths, eighths, and finally, hundredths, when a whole army arrives. Each new situation in the well-written text includes graphic models: "'Three gators,' gulped Alice, 'That means three pieces.' 'Cut three one-thirds,' muttered Al," as a circular model of the pie is shown cut into three equal pieces. The detailed illustrations of the main characters and other woodland creatures humorously complement the story line.

McMillan, Bruce. *Eating Fractions.* New York: Scholastic, 1991.

ISBN 0-590-43770-4 $14.95
ISBN 0-590-72732-X   19.95

★★ Single concept, PS–2

Wholes, halves, thirds, and fourths are the number concepts simply introduced in color photographs. A close-up of one whole banana is followed by a sequence of pictures showing the banana cut into two pieces and the halves being shared between two friends. The simple text includes both the word name and the numeral; for the banana, whole, 1, halves, and 1/2 are introduced next to drawings outlining the whole and the halves. A whole muffin is shown being divided into three pieces; the photographs and the diagrams clearly show three thirds making a whole. All the examples are food, as the title suggests. The author's recipes for the rolls, pizza, salad, and pie appear at the end of the book, which encourages the reader to participate in extension activities that use the concepts explored. The two friends reflect a multicultural setting—one child is white; one is African American.

———. *One, Two, One Pair!* New York: Scholastic, 1991.

ISBN 0-590-43767-4 $12.95

☆ Single concept, PS–2

Socks, shoes, shoelaces, mittens, and ice skates are some of the pairs introduced in this photographic essay reflecting winter scenes. Each object in a pair is shown in a separate photograph on the left-hand page; a larger photograph on the opposite page shows the pair. The minimal text repeated throughout the book is simply "one, two, one pair." Patterns can be detected, as in the sequence from bare feet to socks to shoes. The colorful photographs of the activities of the twins, Jessica and Monica, are appealing. This simple concept book is an effective introduction to the concept of pairs.

Merriam, Eve. *Twelve Ways to get Eleven.* Illustrated by Bernie Karlin. New York: Simon & Schuster, 1993.

ISBN 0-671-75544-7 $14.00

★★★ Single concept, PS–1 (PS–2)

Large, black block numerals from 1 through 10 and 12 are printed across the top of an opening page. But why is 11 missing from the sequence? A green numeral 11 is seen at the base of the page; a turtle is comfortably resting on top of the numeral and smiling at a friendly looking pig. Printed boldly on the next page are "ONE, TWO, THREE, ..., TEN, _ _ _ _, TWELVE" with the query "WHERE'S ELEVEN?" The remainder of the book answers this question in twelve ways. Some examples include nine pinecones and two acorns; six peanuts and five popcorn kernals; four banners, five rabbits, a water pitcher, and a bouquet of flowers; one sow and ten piglets. Each double-page layout is a new surprise, with a simple text listing each new collection that adds to eleven. Karlin's pleasing illustrations were done in cut paper and colored pencils. This is a quality book with a very worthwhile mathematics task for children to pursue.

Moore, Inga. *Six-Dinner Sid.* New York: Simon & Schuster Books for Young Readers, 1991.

ISBN 0-671-73199-8 $13.95
ISBN 0-671-79613-5   (paper)   5.95

★★★ Single concept, PS–3

Six-Dinner Sid is an intelligent cat who has found a way to support a very interesting lifestyle. He lives at six different addresses, numbers 1, 2, 3, 4, 5, and 6 on Aristotle Street. Since none of the people at these addresses knows the others, Sid has six homes, six different personalities to respond to six different owners, six beds, six luxurious ways to be scratched, and six meals a day. Life is perfect for Sid until he becomes ill and ends up at the vet six times in one day. His secret is revealed, and he is relegated to *one* meal a day ... until he moves Pythagoras Place. You've got it! A move to Pythagoras Place, numbers 1, 2, 3, 4, 5, and 6! Moore's unique story is complemented by the lovely artwork, which successfully captures Sid's six personalities, his delicious enjoyment in being scratched, and his disdain for the one paltry meal he receives when discovered. The gentle warmth in color and line provide a secure setting for this resourceful cat. Mathematically, the story line offers the opportunity to look at number uses, both ordinal and cardinal.

Murphy, Stuart J. *Betcha!* Illustrated by S. D. Schindler. New York: HarperCollins Publishers, 1997.

See Series and Other Resources: MathStart

———. *Give Me Half!* Illustrated by G. Brian Karas. New York: HarperCollins Publishers, 1996.

See Series and Other Resources: MathStart

My First Number Book. New York: Dorling Kindersley, 1992.

ISBN 1-879431-73-4 $12.95

★★ Multiconcept, PS–2

Various concepts involving number are introduced in this oversize primary book. Eclectic collections, such as animals, toys, vegetables, tools, and children of different ethnic groups, have been artfully arranged in color photographs against a white background. Classification and seriation tasks are included as well as counting, measuring, and other number tasks. Examples of matched pairs are two shoes or two handprints, whereas matched partners are a racquet and a ball or a knife and a fork. Classification tasks involve questions on how objects are related and how they could be sorted. One-to-one correspondence is used in comparing a set of five puppies to a set of doghouses, bowls, or collars. Sequences are introduced with beads, blocks, and tiles arranged in different patterns for the reader to extend. Different arrangements of groups of children, blocks, or birthday candles illustrate the concept that the number in the group is the same no matter what the arrangement of the objects. One activity in the book uses a six-by-five grid; the object is to match the numerals from 1 through 10 with the appropriate number of both circular counters and garden items. Another includes a board game that requires counting up to fifty. Good opportunities to count and develop number sense appear throughout the book. One double page illustrates the numbers from one through twenty with the appropriate numeral and word names; the objects to be counted are clearly arranged sets of different vegetables. A final page shows the numbers twenty, fifty, one hundred, five hundred, and one thousand. The counters are clearly arranged in arrays; different-colored objects are used so that a rainbow appears in the array of five hundred counters and a clown's face in the array of one thousand.

Ordering tasks encompass both ordering numerals and ordering sets of objects by number. Ordinal concepts from first through tenth are illustrated. Connections to how numbers are used in our world include examples of a clock, bar code, price tag, timer, license plate, and telephone. Addition and subtraction examples are briefly visited with questions, counters, and number sentences. Six pages involve such measurement concepts as comparing size with vocabulary like *thick* and *thin, wide* and *narrow,* and *big* and *small;* measuring length with nonstandard units; and exploring weight. Two pages briefly introduce two- and three-dimensional shapes and their properties and are followed by two pages on time concepts. Questions are posed throughout the book; some are to help connect number to the reader's world; others are to check the reader's understanding of the concepts presented in the book. Because so many different tasks are posed, this book is meant to be sampled over time. Another interesting detail is that each of the forty-eight numbered pages shows a row of counters along the bottom edge next to the appropriate numeral. Other special features of this book are a table of contents, a note to parents, and a glossary.

O'Halloran, Tim. *Know Your Numbers.* Niagara Falls, N.Y.: Durkin Hayes Publishing, 1983.

ISBN 0-88625-045-5 $10.95

★★ Multiconcept, PS–2

Action-packed, humorous, detailed illustrations make this a book to pore over time and time again. Each look will bring another surprise. The rhyming text takes the reader on an exploration of one through twenty in zany situations featuring tinges of subtle humor. An opportunity to test counting skills comes first in the unlikely situation of an expanding moose band and second in a wild scene of monkey firefighters responding to a call. In two situations, the reader is asked to compare pictures to determine whether the rearrangement of objects affects the number of objects. The text then challenges children to use counting or one-to-one correspondence to discern whether there are too many, too few, or just enough to go around in a variety of funny situations featuring rattles for babies, bedrolls for campers, and pies for pigs. Ordinal numbers, simple addition and subtraction, measurable attributes, shapes, and the measurement of time are all presented in the same humorous way but not in enough depth to warrant cross-referencing.

Pinczes, Elinor J. *Arctic Fives Arrive.* Illustrated by Holly Berry. Boston: Houghton Mifflin Co., 1996.
See Early Number Concepts: Counting—Other Number Concepts

Pittman, Helena Clare. *Counting Jennie.* Minneapolis, Minn.: Carolrhoda Books, 1994.
See Number—Extensions and Connections: Statistics and Probability

Pluckrose, Henry. *Numbers.* Illustrated by Chris Fairclough. Chicago: Children's Press, 1995.
See Series and Other Resources: Math Counts

Ryan, Pam Muñoz, and **Jerry Pallotta.** *The Crayon Counting Book.*
Illustrated by Frank Pallotta, Jr. New York: Charlesbridge, 1996.
See Early Number Concepts: Counting—Other Number Concepts

Salt, Jane. *My Giant Word and Number Book.* Illustrated by Sarah
Pooley. New York: Kingfisher Books, 1992.
See Early Number Concepts: Counting—Other Number Concepts

Sandburg, Carl. *Arithmetic.* Illustrated by Ted Rand. San Diego, Calif.:
Harcourt Brace Jovanovich Publishers, 1993.
See Geometry and Spatial Sense: Incidental Geometry—Distortions and
Illusions

Schwartz, David M. *How Much Is a Million?* Illustrated by Steven
Kellogg. New York: Lothrop, Lee & Shepherd Books, 1985.

ISBN 0-688-04049-7		$15.00
ISBN 0-688-04050-0	(library binding)	14.88
ISBN 0-688-09933-5	(paper, Mulberry)	4.95

★★★ Single concept, K–5

Marvelosissimo the Mathematical Magician introduces his eight
friends (four children and four animals) to the magnitude of large
numbers through thought-provoking situations. The detailed pastel
drawings clearly complement the clever fantasy examples used to
introduce 1 million, 1 billion, and 1 trillion. Four different situations
illustrate the enormity of 1 million: finding how high a column of 1
million children standing on each other's shoulders would reach,
finding how long it would take to count to a million, finding a bowl
large enough to hold a million goldfish, and finding how many pages
in the book would be needed to hold a million stars. These examples
are modified and extended to show 1 billion and 1 trillion. Of
particular interest is a three-page note from the author that explains
how he generated the data used in the examples. Middle-grades
children will enjoy the challenge of estimation and the calculation of
similar activities for numbers of their own choice. This excellent book
can be appreciated by readers of any age. A companion teacher's
guide is available from the publisher.

Scieszka, Jon. *Math Curse.* Illustrated by Lane Smith. New York:
Viking, 1995.

ISBN 0-670-86194-4 $16.99

★★★ Multiconcept, 1+ (all ages)

When the teacher, Mrs. Fibonacci, says, "You know, you can think
of almost everything as a math problem," the author and illustrator
take her words to heart! The dedications in this book are posed as
mathematics problems; the list of prior books published by the
author or the illustrator is presented in a Venn diagram. This

uniqueness sets the stage for the story of one day in the life of a young student who is "cursed" as she begins to find mathematics problems in everything she sees and does! The teacher's opening statement becomes the MATH curse! The alarm clock's ringing at 7:15 becomes a problem as she worries whether she has enough time to catch the 8:00 bus, since it takes "10 minutes to dress, 15 minutes to eat my breakfast, and 1 minute to brush my teeth." Eating breakfast quickly generates problems about quarts in a gallon, pints in a quart, feet in a yard, and the number of cereal flakes in a bowl. In school, the study of English becomes a word problem: "If mail + box = mailbox, Does lipstick − stick = lip? Does tunafish + tunafish = fournafish?"

Poor child! Her anxiety grows as she sees mathematics problems, some real, some unreal, all around. The text uncovers a myriad of problems that must be read carefully for the humor embedded within them. After a truly terrifying dream, the young student calms down as she realizes that the problems have solutions! Just as she is enjoying this realization she is faced with a science teacher, Mr. Newton, who says, "You know, you can think of almost everything as a science experiment...." The artwork is a wonderful complement to this amusing and informative text. Changing expressions of innocence, horror, despair, sheer terror, and final self-satisfied bliss are captured in the simple figure of the main character. Also, the growing confusion of things real and fantastic, which jostle for position, is clearly visible on each busy page.

Sis, Peter. *Going Up! A Color Counting Book.* New York: Greenwillow Books, 1989.

ISBN 0-688-08125-8	$12.95
ISBN 0-688-08127-6 (library binding)	12.88

☆ Single concept, PS+

Against the gray pen-and-ink illustration of a city skyline, watercolor paints focus the reader's attention and present an element of intrigue. The opening double-page layout shows a young girl in front of a flower shop. Nearby, all the floors of a twelve-story hotel, except the bottom and top floors, are colorfully lighted, and the text warns mysteriously that everyone is getting ready. Mary takes the elevator from the first floor and is joined at each subsequent floor by an unusually costumed person. Everyone gets off at the twelfth floor, where the mystery is solved—a surprise birthday party is revealed. The final scene in the story shows the gray pen-and-ink illustration of the city skyline, but this time the only floor in the building with bright lights is the twelfth! Although the story itself is one that young children will enjoy, each colorful layout shows only the increasingly crowded elevator. The reference to the main floor as the first, the one where the series begins, is lost. With adult direction the reader could reexplore the concepts of first, second, and so on, by discussing what is depicted on the end pages. Children could ask their own questions to check their classmates' understanding of ordinal numbers in a review of the story.

Stienecker, David L. *Fractions.* Illustrated by Richard Maccabe. New York: Benchmark Books, 1996.
See Series and Other Resources: Discovering Math

Smoothey, Marion. *Calculators.* Illustrated by Ann Baum. New York: Marshall Cavendish, 1995.
See Series and Other Resources: Let's Investigate

———. *Number Patterns.* Illustrated by Ted Evans. New York: Marshall Cavendish, 1993.
See Series and Other Resources: Let's Investigate

———. *Numbers.* Illustrated by Ted Evans. New York: Marshall Cavendish, 1993.
See Series and Other Resources: Let's Investigate

Time-Life for Children Staff. *Alice in Numberland: Fantasy Math.* Alexandria, Va.: Time-Life for Children, 1993.
See Series and Other Resources: I Love Math

———. *The Case of the Missing Zebra Stripes: Zoo Math.* Alexandria, Va.: Time-Life for Children, 1992.
See Series and Other Resources: I Love Math

———. *Look Both Ways: City Math.* Alexandria, Va.: Time-Life for Children, 1992.
See Series and Other Resources: I Love Math

———. *The Mystery of the Sunken Treasury: Sea Math.* Alexandria, Va.: Time-Life for Children, 1993.
See Series and Other Resources: I Love Math

———. *The Search for the Mystery Planet: Space Math.* Alexandria, Va.: Time-Life for Children, 1993.
See Series and Other Resources: I Love Math

Vorderman, Carol. *How Math Works.* Pleasantville, N.Y.: Reader's Digest Association, 1996.
See Series and Other Resources: Resources

Wakefield, Ali. *Those Calculating Crows!* Illustrated by Christy Hale. New York: Simon & Schuster Books for Young Readers, 1996.
See Early Number Concepts: Counting—Other Number Concepts

Walpole, Brenda. *Counting.* Illustrated by Dennis Tinkler and Chris Fairclough. Milwaukee, Wis.: Gareth Stevens, 1995.

See Series and Other Resources: Measure Up with Science

Wells, Robert E. *Is a Blue Whale the Biggest Thing There Is?* Morton Grove, Ill.: Albert Whitman & Co., 1993.

See Measurement: Measurement Concepts and Systems

Number Systems

Are you about 1100 years old? This question was posed to seventh graders in a small town in southern Illinois. To furnish information to help the students answer the question, Watson's *Binary Numbers* was read. The students soon were predicting and estimating the number of pieces of grain and the lengths of a string as they imagined the repeated doubling of both these quantities. One activity with Cuisenaire rods was acted out. Within ten minutes of the initial question, the students had cracked the code of base-two numeration and were busy exploring how to write other base-ten numbers in this format.

The description above highlights Watson's *Binary Numbers,* an exceptional but out-of-print book designed for exploring numeration systems.

Friedman's *The King's Commissioners* provides an opportunity for students to develop a better understanding of their own numeration system as they explore other bases. Historical connections are described in Fisher's *Number Art: Thirteen 1 2 3s from around the World* and Massin's *Fun with 9umbers.* Roman numerals are introduced by both Adler and Geisert. Some narratives convey the evolution of number and numeration systems and the role of zero. The books cross-referenced in this category contain brief, but quality, references to number systems and place value.

The following books are out of print but are considered exceptional:

Adler, David A. *Base Five.*

Sitomer, Mindel, *and* **Harry Sitomer.** *How Did Numbers Begin?*

Watson, Clyde. *Binary Numbers*

Weiss, Malcolm E. *Solomon Grundy, Born on Oneday: A Finite Arithmetic Puzzle*

Zaslavsky, Claudia. *Zero: Is It Something? Is It Nothing?*

Copies may be available in your library. See the first edition of *The Wonderful World of Mathematics* for annotations.

Adler, David A. *Roman Numerals.* Illustrated by Byron Barton. New York: Thomas Y. Crowell Publishers, 1977.

ISBN 0-690-01302-7 $14.89

★★★ Single concept, K–3 (2–4)

Children of any age are encouraged to accept this compelling invitation to explore the meaning and use of roman numerals. The invitation comes from a thoughtful Roman stonecutter who presents a conversational, but clear and concise, explanation of how the roman numeration system works. The invitation to think through, and participate in, a variety of activities becomes more tempting as the illustrations involve the reader in a logical presentation of what is described in the text.

Fisher, Leonard Everett. *Number Art: Thirteen 1 2 3s from around the World.* New York: Four Winds Press, 1982.

ISBN 0-590-071810-0 $12.95

★★ Single concept, 3–7

The history and development of thirteen number systems are presented: Arabic, Armenian, Brahman, Chinese, Egyptian, Gothic, Greek, Mayan, Roman, runic, Sanskrit, Thai, and Tibetan. As each number system is given, its history, cultural impact, development, and use are discussed. Additionally, the actual symbols (numerals, letters, or characters) of the system are included. Within the descriptions, the author includes cross-references to the influences of other cultures discussed in the book. The reader should appreciate the gradual evolution of our current system and the interrelatedness of our world cultures. The book is visually pleasing; graceful illustrations in blue and white complement this informative and interesting text.

Friedman, Aileen. *The King's Commissioners.* Illustrated by Susan Guevara. A Marilyn Burns Brainy Day Book. New York: Scholastic, 1995.

ISBN 0-590-48989-5 $14.95

★★★ Single concept, 1–4

Appointments by the King include the Commissioner for Flat Tires, the Commissioner for Golf Balls, the Commissioner for Late Arrivals, the Commissioner for Mismatched Socks, the Commissioner of Spilt Milk, and the Commissioner for Lost Homework. Since these were just a *few* of the King's commissioners, it became the duty of the First and Second Royal Advisors to count all of them. As each commissioner walked through the throne room door, each advisor marked a tally. One advisor grouped his tallies by twos and reported

23 twos and one more; the other advisor grouped by fives and reported 9 fives and two more. The King becomes impatient because he wanted *a* count. The Princess resolved the problem by having all the advisors line up in rows of ten. She explained how forty-seven is related to both reports by the Royal Advisors. Hilarious text and bold, vigorously painted illustrations convey the humor of the situation and expand its meaning to stimulate children's thinking about grouping and our number system. A worthwhile mathematical task is embedded in this exceptional book. Two pages of notes for parents, teachers, and other adults guide them through a better understanding of the mathematics in the book, of how children learn mathematics, and of how they might discuss the mathematics in this book with young readers.

Geisert, Arthur. *Roman Numerals I to MM = Numerabilia Romana Uno ad Duo Mila.* Boston: Houghton Mifflin, 1996.

ISBN 0-395-74519-5 $15.95

★★ Single concept, (2–4)

Literally thousands of pigs are featured in Geisert's intricate etchings that introduce the Roman numeral system. To determine the value of each of the seven letters used in the system, the reader counts the number of pigs in etchings labeled I, V, X, L, C, D, and M. L pigs climb all over a jungle gym, C pigs frolic in a playground with three teeter-totters, D pigs play in and around a pond, and M pigs are in a large pasture. After introducing the basic symbols, the illustrations and text clearly describe the multiple meanings of the symbols, such as III, XX, and CCC. Using both addition and subtraction to write numbers is illustrated by examples of the first twenty counting numbers. The remaining pages are counting puzzles. At the bottom of each page a list of objects to be counted is given; for example, in one park scene the objects to be counted include IX flowerpots, XVI gopher holes, IV stone posts, III trash cans, and VL pigs. The illustrations are humorously done with pigs sliding down slides, digging in the mud, engaging in a tug of war, or lying on their backs. The author claims that MMMMDCCCLXIV pigs are pictured in the book; readers will not challenge his count.

Griffiths, Rose. *Numbers.* Illustrated by Peter Millard. Milwaukee: Gareth Stevens Publishing, 1994.

See Series and Other Resources: First Step Math

Massin. *Fun with 9umbers.* Illustrated by Les Chats Pelés. San Diego, Calif.: Creative Editions, 1993.

ISBN 0-15-200962-0 $17.00

★★ Multiconcept, 2–7 (3–7)

Part of mathematical literacy is developing an appreciation of our history, and Massin gives students such an opportunity in *Fun with*

9umbers. Numeration systems based on numbers other than ten are encountered. Noting that the Aztecs of Mexico used twenty as a base because they were barefoot and could readily use fingers *and* toes for counting provides the human dimension to the evolution of numeration systems. The Egyptian symbols of a coiled rope for 100 and an astonished man for 1 000 000 can readily generate a discussion on the historical origins of the numerals in our system. The origins of zero and its spread across cultures are noted. The historical origins of metric and customary systems of measurement are also introduced in terms of the human dimension. Surprising connections across cultures include Julius Caesar's decree of 365 days in a year and the number of steps to the top of a Mayan pyramid. The continuing evolution and use of number and numerals are noted in the influence of the calculator. Stylized numerals are cleverly and whimsically incorporated into the sophisticated graphic illustrations created by Les Chats Pelés, a trio of artists: Lionel Le Néouanic, Benoît Morel, and Christian Olivier.

Scieszka, Jon. *Math Curse.* Illustrated by Lane Smith. New York: Viking, 1995.

See Number—Extensions and Connections: Number Concepts

Stienecker, David L. *Numbers.* Illustrated by Richard Maccabe. New York: Benchmark Books, 1996.

See Series and Other Resources: Discovering Math

Time-Life for Children Staff. *From Head to Toe: Body Math.* Alexandria, Va.: Time-Life for Children, 1993.

See Series and Other Resources: I Love Math

———. *See You Later Escalator: Mall Math.* Alexandria, Va.: Time-Life for Children, 1993.

See Series and Other Resources: I Love Math

Vorderman, Carol. *How Math Works.* Pleasantville, N.Y.: Reader's Digest Association, 1996.

See Series and Other Resources: Resources

Walpole, Brenda. *Counting.* Illustrated by Dennis Tinkler and Chris Fairclough. Milwaukee: Gareth Stevens, 1995.

See Series and Other Resources: Measure Up with Science

Operations

As part of an exploration of multiplication, Corrine Brown read Pinczes's *One Hundred Hungry Ants* to the children in her multiage classroom. To extend their investigation, each child selected a number from 1 through 200 to create a personal book about 100 hungry ants. One third grader selected 15 and diligently worked to arrange fifteen color tiles into various rectangles; after much work, he discovered that 1, 3, 5, and 15 were factors. Mitchell, another third grader, selected the number 200 and worked with base-ten blocks. He wondered if the ants could sit on one another's shoulders, and created a $4 \times 5 \times 10$ rectangular prism. Although the boys were at different levels mathematically, they both successfully communicated their work and were proud of their accomplishments.

More than the books in any other category, those in "Operations" have evolved in quality and content. A wider selection of excellent mathematical tasks is now represented. Developing and analyzing the concept of an operation make up one aspect of this category. Butler's *Too Many Eggs,* Mathews's *Bunches and Bunches of Bunnies,* and Moerbeck and Dijs's *Six Brave Explorers* use fantasy stories and illustrations to introduce operation concepts; Hutchins's approach in *The Doorbell Rang* is to connect the concepts to "real life" situations. In many books, the reader is to count or act out the situation using models to solve the problems; in Butler's *Too Many Eggs,* counters are included for the reader to act out the story. Number sentences are included in some books, such as those by Chorao and Murphy.

Several operations are explored in the lively illustrations by Chorao and Jonas. The unique approach and stunning illustrations found in Edens's *How Many Bears?* invite the reader to analyze number relationships and operations. Doubling is introduced in Hong's *Two of Everything: A Chinese Folktale.* Division and multiplication properties and patterns are extended in books like Chwast's *The Twelve Circus Rings* and Pinczes's *A Remainder of One* and *One Hundred Hungry Ants.* Many books in this category could lead children to write or tell their own stories and share them with their classmates.

Zimelman's *How the Second Grade Got $8205.50 to Visit the Statue of Liberty* is a delightful combination of fantasy and real life that involves addition and subtraction. Other applications and explorations that involve number and operation sense can be found on some pages in the I Love Math series by the Time-Life for Children Staff. Work with large numbers and algorithms appears in the Discovering Math series by Stienecker and Wells. Operations using larger numbers are sometimes embedded in trick problems leading to number theory ideas. Some books in this category include mental computation and estimation.

The following books are out of print but are considered exceptional:

Froman, Robert. *The Greatest Guessing Game: A Book about Dividing*

Mathews, Louise. *The Great Take-Away*

Copies may be available in your library. See the first edition of *The Wonderful World of Mathematics* for annotations.

Adler, David A. *Calculator Riddles.* Illustrated by Cynthia Fisher. New York: Holiday House, 1995.

ISBN 0-8234-1186-9	$12.95
ISBN 0-8234-1269-5 (paper)	4.95

☆ Single concept, PS–3 (2–5)

Each of the forty-five riddles can be answered by keying in the given number expression on a calculator and reading the displayed answer upside down. Typical of the riddles are "Who invented the telephone, but had no one to call? Answer: $193 \times 20 + 9 \times 2 =$ " (BELL) or "What can change an ear into a bear? Answer: $109 \times 48 \div 6 \div 109 =$ " (B). It should be noted that the expressions are to be entered as a string in the order given. The order of operations is ignored and if used would result in an incorrect answer for some riddles. This inconsistency could readily have been avoided if each chain had been written as a series of keystrokes rather than as a numerical expression. Both the numerical and word answers are listed in the back of the book. Black-and-white illustrations enliven or add humor to each riddle. *Calculator Riddles* might motivate readers to write their own riddles.

Anno, Masaichiro, and **Mitsumasa Anno.** *Anno's Mysterious Multiplying Jar.* Illustrated by Mitsumasa Anno. New York: Philomel Books, 1983.

See Number—Extensions and Connections: Number Theory, Patterns, and Relationships

Axelrod, Amy. *Pigs Will Be Pigs.* Illustrated by Sharon McGinley-Nally. New York: Simon & Schuster Books for Young Readers, 1994.

See Measurement: Money

Barry, David. *The Rajah's Rice: A Mathematical Folktale from India.* Illustrated by Donna Perrone. New York: Scientific American Books for Young Readers, 1994.

See Number Theory, Patterns, and Relationships

Brown, Marc Tolon. *Arthur's TV Trouble.* Boston: Little, Brown & Co., 1995.

See Measurement: Money

Butler, M. Christina. *Too Many Eggs.* Illustrated by Meg Rutherford. Boston: David R. Godine Publisher, 1988.

ISBN 0-87923-741-4 $15.95

★★★ Single concept, PS–2

   Pastel illustrations enhance this charming story of Mrs. Bear, who is baking a cake for Mr. Bear's birthday but does not know how to count. The reader helps Mrs. Bear by taking punched-out eggs from the cupboard in the back of the book and placing them in Mrs. Bear's mixing bowls throughout the story. Mrs. Bear believes that she used only six eggs in her cake, but the reader is invited to collect the eggs from Mrs. Bear's mixing bowls and count them to find out how many were actually used. The story line and design of the book are unique. Together they provide a delightful way to extend children's counting activities to beginning addition.

Calmenson, Stephanie. *Dinner at the Panda Palace.* Illustrated by Nadine Bernard Westcott. New York: Scholastic, 1991.

See Early Number Concepts: Counting—Operations

Carle, Eric. *Draw Me a Star.* New York: Philomel Books, 1992.

See Geometry and Spatial Sense: Shapes

Chorao, Kay. *Number One Number Fun.* New York: Holiday House, 1995.

ISBN 0-8234-1142-7 $15.95

★★ Single concept, PS–3 (K–2)

   Addition and subtraction problems are hurled at the reader by Ringmaster Rat, as animals tumble, prance, pile up, fall down, balance, wobble, and squabble their way through this humorous, busy text! The debonair ringmaster appearing on the left-hand pages introduces his unlikely, and unskilled, circus performers while carrying a placard that anticipates the number problem he is going to pose for the reader on each double page. Each time, the answer can be found by considering the antics of the featured performers. The numerical answer to the problem is written on a floating balloon anchored by a little rat who is always on the right-hand page. Pigs piled in a pyramid represent the problem 4 + 3 + 2 + 1; the illustration cleverly depicts this sequence while incidentally introducing the triangular number ten. The problems become progressively more difficult, especially when the types of performers are mixed, as in "Five tumbling mice join four circling cats. / How many mice and cats wearing hats?" In addition, extra challenge is provided through the busyness and detail of the illustrations, which sometimes makes it difficult for readers to find what they are looking for. The final page reviews the number sentences representing the challenges that Ringmaster Rat posed to the reader throughout the lively text.

Chwast, Seymour. *The Twelve Circus Rings.* San Diego, Calif.: Gulliver Books, Harcourt Brace Jovanovich, 1993.

ISBN 0-15-200627-3 $14.95

★★★ Single concept, PS–3 (K–4)

"In the first circus ring, my sister saw with me a daredevil on a high wire." Prose that parallels "The Twelve Days of Christmas" introduces the reader to the action in each of the twelve circus rings. The performers accumulate, with "six acrobats, five dogs a-barking, four aerialists zooming, three monkeys playing, two elephants, and a daredevil on a highwire" in the sixth circus ring. The color, excitement, and busyness of the twelfth ring make it a sight to behold! At the book's end, questions are posed, for example, How many aerialists are zooming? To answer this question, the reader needs to consider that the four aerialists appear in rings four through twelve. Answers and the reasoning behind the answers are clearly stated. Additionally, the author includes other extensions involving patterns. The colorful illustrations capture the many facets and excitement of the circus. As characters accumulate, the reader will pour over the details in each illustration. One surprise includes the unique numbering pattern on the spectators' chairs.

Demi. *One Grain of Rice: A Mathematical Folktale.* New York: Scholastic Press, 1997.

See Number—Extensions and Connections: Number Theory, Patterns, and Relationships

Edens, Cooper. *How Many Bears?* Illustrated by Marjett Schille. New York: Atheneum, 1994.

ISBN 0-689-31923-1 $14.95

★★★ Single concept, K–3

Intriguing hints and fanciful illustrations are highlights in this unique number book. On the opening page, the challenge for the

reader is to discover how many bears run the bakery in Little Animal
Town. A clue is given as each shop in the town is visited. By counting
the number of animals on the opposite page, the reader can
determine the number of bears. For example "In Little Animal Town
... it takes four fewer Giraffes to run the Soda Fountain than it takes
to run the Bakery." Four miniature giraffes are pictured standing on
a soda fountain laden with such typical old-fashioned fare as a
sundae, an ice cream soda, and aluminum containers for straws and
toppings. Another scene shows twenty-four tiny hippopotamuses on a
counter full of containers of fresh and potted flowers; the caption on
the opposite page says, "four times as many hippos to run the Flower
Shop." Readers will explore operations and number as they decide
how to answer each clue. A final page shows the number of animals
in each shop, with a banner proclaiming the numeral. The exquisite
illustrations and the format for this book are a delight, and children
will enjoy the opportunity to explore counting and operations.

Giganti, Paul, Jr. *Each Orange Had Eight Slices: A Counting Book.*
Illustrated by Donald Crews. New York: Greenwillow Books, 1992.

ISBN 0-688-10428-2 $15.00
ISBN 0-688-10429-0   (library binding)   14.93

★★★ Single concept, PS+ (K–3)

Taking an idea from the first line of a familiar poem, "As I was
going to St. Ives," Crews and Giganti expand on it to emphasize that
mathematics concepts and problems can be found everywhere.
Multiplication concepts are explored throughout this unique number
book. In the first encounter, 3 red flowers are seen; each flower has 6
petals and each petal has 2 black bugs. Questions are posed about
the number of flowers, the number of petals, and the number of bugs.
Each problem is posed in parallel format, but the tasks vary; for
some problems, addition and multiplication are involved. This clever
shift demands the reader's full attention. Crew's illustrations
complement the simple text; uncluttered drawings clearly show the
groups and objects to be considered. Young children may solve the
problems with counting or repeated addition. Their final challenge is
to respond to the original poem on which this story is based. How
many were going to St. Ives? Giganti and Crews have designed an
effective format for introducing multiplication that children should
find exciting and be eager to emulate.

Griffiths, Rose. *Number Puzzles.* Illustrated by Peter Millard.
Milwaukee, Wis.: Gareth Stevens Publishing, 1995.

See Series and Other Resources: First Step Math

Hong, Lily Toy. *Two of Everything: A Chinese Folktale.* Morton Grove,
Ill.: Albert Whitman & Co., 1993.

ISBN 0-8075-8157-7 $14.95

★★ Single concept, K–3

Muted, dark tones in blues, greens, and browns effectively illustrate the retelling of this Chinese folktale. Through the use of color and outline, Hong has created flat, but pleasing, illustrations. A poor old farmer finds a brass pot that he takes home to his wife. They soon realize that the pot is magic when they discover in the pot two coin purses containing five gold coins although they had hidden only one coin purse in it. The doubling magic of the pot becomes the couple's good fortune, but as in most tales of magic, trouble follows. The old farmer accidentally discovers an appropriate solution to the problem that develops. This charming folktale introduces the reader not only to another culture but also to the concept of doubling.

Hulme, Joy N. *Counting by Kangaroos.* Illustrated by Betsy Scheld. New York: W. H. Freeman & Co., 1995.

ISBN 0-7167-6602-7 $15.95

★★ Single concept, (1–3)

Kangaroos, wombats, quokkas, numbats, and bandicoots are some of the animals used to illustrate multiplication by three in this story set in Australia. Skip counting is included in the rhyming text for counting the total in each of three equal sets. The three kangaroos that hippity hop up to Sue and Fay's door are the basis for multiplying by three. Once inside the house, they shed one hat each (1, 2, 3) and a pair of shoes each (2, 4, 6). Inside each kangaroo's pouch are "Critters all cramped / and crowded inside, / Australian animals hitching a ride." Groups of three squirrel gliders (3, 6, 9), four koalas (4, 8, 12), five bandicoots (5, 10, 15), ..., nine echidnas (9, 18, 27), and ten wallaby joeys (10, 20, 30) crowd the house. Overrun by all the animals, Sue starts a zoo. The illustrations have a pleasing, primitive, childlike quality.

————. *Sea Squares.* Illustrated by Carol Schwartz. New York: Hyperion Books for Children, 1991.

ISBN 1-56282-079-6 $13.95
ISBN 1-56282-080-X   (library binding)   13.89
ISBN 1-56282-520-8   (paper)   4.95

★★ Single concept, PS–3 (1–3)

Eye-catching illustrations and romping rhyme invite the reader to "Come with me to the side of the sea, / Where the ocean meets the shore. / We'll count some creatures that crawl and creep / Or grow on the ocean floor." Each double-page spread is more appealing than the previous one. The combination of color, clarity, and focus in the illustrations and the alliterative, humorous text is very satisfying: "8 'octos' on the ocean floor, / have scrambled legs, 64." Beyond a counting book, this text introduces squaring numbers and enhances number sense as the reader views the representations and relative sizes of seven sets of seven, eight sets of eight, and so on. Each left-hand page introduces from one through ten creatures with the number word embedded in the poem but written in bold italics. *"Seven* heavy pelicans diving for their dinner. / Seven fish in every pouch can never make them thinner." The challenge in the text on the right-hand page is to count and check the squared number given, for example, "7 pouchy pelicans / Gulp 49 fish with fins." Although the square numbers are illustrated as sets, the reader may wish to represent these numbers with arrays to discover why the book is called *Sea Squares*. An interesting visual challenge is given in the illustrations where the contents of the right-hand border invite the reader to predict what comes next. A different variety of shells is featured in each left-hand border. A beautiful coastal sunset draws this refreshing book to a satisfying close. A factual and more detailed description of each featured animal is included for those who may be motivated to learn more.

————. *Sea Sums*. Illustrated by Carol Schwartz. New York: Hyperion Books for Children, 1996.

ISBN 0-7868-0170-0 $13.95
ISBN 0-7868-2142-6   (library binding)   13.89

☆ Single concept, PS–3 (K–2)

As in Hulme's previous books, the problems posed are in the context of the ocean and its inhabitants. Yellow and green striped triggerfish, spiky urchins, bristling lionfish, frigate birds, and swaying sponges are some of the colorful sea creatures that grace the attractive illustrations. Both addition and subtraction situations, either one step or multistep, are posed in the text with a number sentence written in the lower right of each double-page spread. The complexity varies from 1 siphon + 1 siphon = 2 siphons to sentences like 24 legs − 12 legs = 12 legs or 10 polyps + 10 polyps + 10 polyps = 30 polyps. Some of the situations are clearly depicted, such as the one with groups of two urchins and three urchins and one more urchin on the opposite page to illustrate 2 + 2 + 1 = 6. In others, the action is not conveyed in the illustrations. For example, two clams and four more clams are shown on the ocean floor with the text explaining that the stingray "chews **three** clams for breakfast, / And crunches **three** for brunch / so all the clams are gobbled up, / and none are left for lunch," which is meant to represent 2 clams + 4 clams − 3 clams − 3 clams = 0 clams. Another layout shows a booby swooping underwater with three of six fish captured in its bill; the text notes that it gulps all six fish and captures two more, which are not shown; the opposite page states that another bird, a frigate, grabs these two fish so the booby has only 6 fish + 2 fish − 2 fish = 6 fish for lunch. Although this book offers an opportunity to introduce young children to story situations and life on a coral reef, the examples are often contrived and confusing. Additional information about the inhabitants of coral-reef communities is included in the closing pages.

Hutchins, Pat. *The Doorbell Rang.* New York: Greenwillow Books, 1986.
ISBN 0-688-05252-5 $15.93
ISBN 0-688-09234-9   (paper, Mulberry)   4.95
★★ Single concept, PS–3

Sharing cookies among friends introduces beginning division concepts. The simple text opens with Victoria and Sam starting to share twelve cookies and comparing them with Grandma's cookies. Before they can place six cookies on each plate, the doorbell rings, and they are joined by two next-door neighbors. The doorbell continues to ring; the children eventually must divide the cookies among six and then twelve. Victoria and Sam's mother welcomes each group as well as the surprise guest at the end of the book— Grandma with more cookies. Each double-page spread of the growing number of children gathered around the kitchen table is colorfully illustrated. *The Doorbell Rang* has become a favorite of teachers and children because it is an excellent springboard for discussing division, remainders, and fractions.

Jonas, Ann. *Splash.* New York: Greenwillow, 1995.
ISBN 0-688-11051-7 $15.00
ISBN 0-688-11052-5   (library binding)   14.93
★★ Single concept, PS+ (PS–1)

"How many are in my pond?" asks the text as one turtle jumps into the pond occupied by two catfish and four goldfish. Scenes from a summer day around the pond provide the playful story line for this pleasant book, which informally introduces addition and subtraction. Each situation is presented through a simple story line and is captured in the colorful illustrations. The turtle jumps into the pond, followed by the frog as the cat watches the fish. The cat and the dog fall in followed by two more frogs; as the waterlogged cat and dog climb out, a dragonfly falls in and a frog hops out. Even the pretty young African American girl who tells the story falls in! After each series of actions, the question is posed, "How many are in the pond?" Jonas has provided a delightful introduction to addition and subtraction situations.

Leedy, Loreen. *2 × 2 = Boo! A Set of Spooky Multiplication Stories.* New York: Holiday House, 1995.
ISBN 0-8234-1190-7 $15.95
☆ Single concept, 3–4 (2–4)

Halloween is the vehicle for introducing the multiplication facts from zero through five. The author cleverly combines a story line with each fact family. Each multiplication family is introduced in a separate chapter with the first one titled "The Disappearing Zero."

After witch Griseda shows two black cats how to make things disappear, they yell out "$3 \times 0 = 0$" at three snakes chasing them; the next panel shows the shadows of the vanished snakes. In "One More Time," combinations from 1×1 through 5×1 are illustrated in Dr. Albert's multiplication machine, which effectively illustrates how one Halloween trick-or-treat bag becomes three bags after the machine is programmed for 3×1. In "Seeing Doubles," magic eyeglasses double everything from 1 through 5 items; however, the number combinations are reversed, since standard notation for two groups of three witch hats would be 2×3 rather than the given 3×2. Similar sentences are illustrated in other chapters; for example, three pumpkins are shown with two eyes each and the situation is expressed as "two eyes times THREE equals six eyes" and written as the number sentence $2 \times 3 = 6$. The fives are appropriately represented as two sets of five, three sets of five, and so on. Summary pages show each family of facts illustrated with objects correctly arranged in arrays. Throughout the book questions about each combination can be answered by counting the objects in the set. Although the families for zero and one make up the first chapters, this order of presentation would not typically be the reader's first introduction to multiplication; conceptually, these families should be introduced only after other multiplication facts have been explored. However, appropriate generalizations for the ones and the zeroes families are illustrated.

Long, Lynette. *Domino Addition.* Watertown, Mass.: Charlesbridge Publishing, 1996.

ISBN 0-88106-878-0		$14.95
ISBN 0-88106-879-9	(library binding)	15.88
ISBN 0-88106-877-2	(paper)	6.95

☆ Single concept, K–3 (K–2)

Bold presentation of black and white on solid colored pages make this book very attractive and pleasing to the eye. Long's purpose is to provide some addition fun through the use of dominoes. Illustrations and page titles dominate; the directions for each page are given in smaller type. On a left-hand page announcing, "The total is TWO!", models of two dominoes are shown with the addition facts $2 + 0 = 2$ and $1 + 1 = 2$ written vertically next to each domino. On the right-hand page, the challenge is to find the dominoes with the same total in an arrangement of seven dominoes representing the shape of the numeral 2. Readers can check their success on the following page where, in the right-hand lower corner, a little box reflects the dominoes' arrangement with the correct ones colored. Basically the reader can determine which dominoes to choose by matching parts or patterns. A final page summarizes domino addition facts for the sums from 0 through 12.

Mathews, Louise. *Bunches and Bunches of Bunnies.* Illustrated by Jeni Bassett. New York: Scholastic, 1978.

ISBN 0-590-44766-1 (paper) $3.95

☆ Single concept, K–3

Bunches of bunnies show multiplication of a number by itself. Each set of bunnies is uniquely drawn and is introduced through humorous rhyming verse. One page of a two-page spread has three sets of three bunnies planting, weeding, and watering. The opposite page shows the nine bunnies enjoying the harvest of their garden. The detailed illustrations depict 1×1 to 12×12 in set format.

McGrath, Barbara Barbieri. *The "M&M's " Brand Chocolate Candies Counting Book.* Watertown, Mass.: Charlesbridge Publishing, 1994.

See Early Number Concepts: Counting—Other Number Concepts

McMillan, Bruce. *Jelly Beans for Sale.* New York: Scholastic Press, 1996.

See Measurement: Money

Moerbeck, Kees, and **Carla Dijs.** *Six Brave Explorers.* Los Angeles: Price Stern Sloan, 1988.

ISBN 0-8431-2253-6 $9.95

★★★ Single concept, PS–3

Six brave explorers have encounters with pop-up creatures: a large bird, a cobra, a panther, a hyena, and a crocodile. In each successive subtraction situation, one explorer is "taken away." The explorers are countable, and subtraction is represented; for example, on the left-hand page, a reader sees six explorers, and on the right-hand page, one of the six explorers is being carried away. The short, rhyming narrative has a surprise ending. This triangular-shaped pop-up storybook is unique in design. Its sturdy pages will withstand frequent use. Vibrant color and the action of the characters capture the mystery of the Nile.

Murphy, Stuart J. *Ready, Set, Hop!* Illustrated by Jon Buller. New York: HarperCollins Publishers, 1996.

See Series and Other Resources: MathStart

———. *Too Many Kangaroo Things to Do!* Illustrated by Kevin O'Malley. New York: HarperCollins Publishers, 1996.

See Series and Other Resources: MathStart

Nayer, Judy. *Funny Bunnies.* Illustrated by Steve Henry. New York: McClanahan Book Company, 1994.

ISBN 1-56293-436-8 $4.95

☆ Single concept, (PS–1)

This sturdy book with movable flaps provides a limited number of addition and subtraction possibilities. On the reverse side of the raised flap, the number sentence representing the story's action appears. Each subsequent situation begins with the same number of bunnies that was the answer from the previous page. For example, the text reads, "Four little bunnies hop to a tree. One climbs up it...." The reader lifts the flap, which hides the one bunny and reveals more text inside: "Now there are three. 4 − 1 = 3. You can subtract!" Three bunnies appear in the next situation and are joined by another to introduce 3 + 1 = 4. The seven situations involve no numbers larger than five. The illustrations are bright and colorful, clearly reflecting the message of the text.

————. *Tricky Puppies*. Illustrated by Steve Henry. New York: McClanahan Book Co., 1994.

ISBN 1-56293-435-X $4.95

☆ Single concept, (PS–1)

Playful puppies introduce seven addition and subtraction situations in the same format as in Nayer's *Funny Bunnies*. For the first five examples, the common thread from one situation to the next is the number of playful puppies. In one example, two playful puppies jumping over the wall effectively illustrate 2 − 2 = 0. Numbers no larger than six are considered. For a more complete description see Nayer's *Funny Bunnies*.

Pinczes, Elinor J. *A Remainder of One*. Illustrated by Bonnie MacKain. Boston: Houghton Mifflin Co., 1995.

ISBN 0-395-69455-8 $14.95

★★★ Single concept, PS–3 (1–4)

Pinczes and MacKain have produced another excellent children's book, parallel in content and design to *One Hundred Hungry Ants*. In this story Joe is one of twenty-five bugs in the marching infantry. When they march for the queen, rows of two are formed and Joe, the remainder of one, is asked to step aside as a honeybee sternly warns, "the queen likes things tidy." For the next parade the bug infantry marches in rows of three, forming three rows of eight; the following day rows of four are formed. Each time Joe is "a remainder of one" and is asked to step aside. That night Joe decides that five rows *must* work. The Queen is proud to see the perfect results, and Joe is happy to be "the former remainder of one." The illustrations, outlined in black, depict meadow scenes in tones of army green, browns, and purples. This exceptional book simply but effectively introduces division concepts. These concepts can be readily extended by changing the number being considered, either the whole or the remainder.

———. *One Hundred Hungry Ants*. Illustrated by Bonnie MacKain. Boston: Houghton Mifflin Co., 1993.

ISBN 0-395-63116-5 $14.95

★★★ Single concept, PS–3

One hundred hungry ants set off in single file to a picnic; they hurry because they are concerned that most of the food will be gone. But as they march, their progress seems too slow, so the littlest ant suggests that they reorganize into two lines of 50. This scenario is repeated as the 100 ants rearrange into 4 lines of 25, 5 lines of 20, and finally 10 lines of 10. "One hundred ants were singing / and marching in 10 rows. / 'At last, we're at the picnic! / A hey and a hi dee ho!'" The lively story, told in rhyming repetitive verse, ends as the ants discover that all the food is gone because they "took so long with rows." The illustrations, outlined in black, depict woodland scenes in tones of army green, browns, and purples. This is an exceptional book that simply, but effectively, introduces the factors of 100.

Scieszka, Jon. *Math Curse*. Illustrated by Lane Smith. New York: Viking, 1995.
See Number—Extensions and Connections: Number Concepts

Smoothey, Marion. *Calculators*. Illustrated by Ann Baum. New York: Marshall Cavendish, 1995.
See Series and Other Resources: Let's Investigate

Stienecker, David L. *Addition*. Illustrated by Richard Maccabe. New York: Benchmark Books, 1996.
See Series and Other Resources: Discovering Math

———. *Division*. Illustrated by Richard Maccabe. New York: Benchmark Books, 1996.
See Series and Other Resources: Discovering Math

———. *Multiplication*. Illustrated by Richard Maccabe. New York: Benchmark Books, 1996.
See Series and Other Resources: Discovering Math

Time-Life for Children Staff. *Alice in Numberland: Fantasy Math*. Alexandria, Va.: Time-Life for Children, 1993.
See Series and Other Resources: I Love Math

———. *The Case of the Missing Zebra Stripes: Zoo Math*. Alexandria, Va.: Time-Life for Children, 1992.
See Series and Other Resources: I Love Math

————. *From Head to Toe: Body Math.* Alexandria, Va.: Time-Life for Children, 1993.
See Series and Other Resources: I Love Math

————. *How Do Octopi Eat Pizza Pie: Pizza Math.* Alexandria, Va.: Time-Life for Children, 1993.
See Series and Other Resources: I Love Math

————. *The Mystery of the Sunken Treasury: Sea Math.* Alexandria, Va.: Time-Life for Children, 1993.
See Series and Other Resources: I Love Math

————. *Play Ball: Sports Math.* Alexandria, Va.: Time-Life for Children, 1993.
See Series and Other Resources: I Love Math

————. *Pterodactyl Tunnel: Amusement Park Math.* Alexandria, Va.: Time-Life for Children, 1993.
See Series and Other Resources: I Love Math

————. *Right in Your Own Backyard: Nature Math.* Alexandria, Va.: Time-Life for Children, 1993.
See Series and Other Resources: I Love Math

————. *The Search for the Mystery Planet: Space Math.* Alexandria, Va.: Time-Life for Children, 1993.
See Series and Other Resources: I Love Math

————. *See You Later Escalator: Mall Math.* Alexandria, Va.: Time-Life for Children, 1993.
See Series and Other Resources: I Love Math

Vorderman, Carol. *How Math Works.* Pleasantville, N.Y.: Reader's Digest Association, 1996.
See Series and Other Resources: Resources

Wells, Alison. *Subtraction.* Illustrated by Richard Maccabe. New York: Benchmark Books, 1996.
See Series and Other Resources: Discovering Math

Zimelman, Nathan. *How the Second Grade Got $8205.50 to Visit the Statue of Liberty.* Illustrated by Bill Slavin. Morton Grove, Ill.: Albert Whitman & Co., 1992.
See Measurement: Money.

Statistics and Probability

Opportunities for collecting, organizing, and describing data can be found in this collection of books on predictions through probability and data analysis. These books describe many excellent activities for constructing and selecting appropriate graphs as well as for reading and interpreting displays of data—tables, charts, and graphs. A number of situations offer opportunities for readers to formulate and solve problems through collecting and analyzing data. Some connections to the real world reflect the children's world; others encompass larger segments of the population and environment. Currently there are fewer books that explicitly relate to the study of probability.

Most of the books in this category are cross-referenced, since they are multiple-concept books or the data are embedded in another mathematical topic. For young children, Pittman's *Counting Jennie* is a playful introduction to using number to describe the world around us. Murphy's *The Best Vacation Ever* considers the question of what data are to be collected and then uses tallies to present the data. The Time-Life for Children Staff has incorporated a simple pictograph or bar graph in one episode in each book. Fanelli's *My Map Book* includes a broad interpretation of different ways to present various types of information. For older readers, the format and intriguing data in Ash's *Incredible Comparisons* should motivate them to reflect on using data, develop intriguing ways to display data, and design "incredible comparisons" of their own. Clement's *Counting on Frank* and Scieszka's *Math Curse* also contain delightful examples that integrate data, measurement, and number. More sophisticated data experiments connected to the real world can be found in Walpole and Markle; concepts of probability are explored in the works by Cushman and Smoothey.

The following books are out of print but are considered exceptional:

Arnold, Caroline. *Charts and Graphs: Fun, Facts, and Activities*

Linn, Charles F. *Probability*
Mori, Tuyosi. *Socrates and the Three Pigs*
Parker, Tom. *In One Day*
Srivastava, Jane Jonas. *Averages*

Copies may be available in your library. See the first edition of *The Wonderful World of Mathematics* for annotations.

Ash, Russell. *Incredible Comparisons.* New York: Dorling Kindersley, 1996.
ISBN 0-7894-1009-5 $19.95
★★★ Single concept, 4–7

Incredible *is* one of the words to describe *Incredible Comparisons*. Fascinating facts about twenty-three topics are graphically displayed in this oversize book; the explanatory text is brief, since the illustrations use scale drawings and other models to convey the relative sizes effectively. One topic, titled "On the Surface," relates facts about the 29 percent of the earth's surface that is land. Different-sized squares are used to compare the size of the continents; the text gives their total area and relates what percent this is of Earth's total land. A quarter of all land is desert, and the areas of five deserts are represented by different-sized squares. Another map shows that the Sahara desert is as large as the continental United States, but the text states that the U.S. population is 130 times as large as that of the Sahara. Maps of the United States, Australia, France, and the Sudan are layered on top of each other to show relative areas; the areas are also stated to the nearest square mile and nearest square kilometer. A line denotes the widest part of each country; the widths are also compared by the amount of time needed to travel that distance with France's TGV, a high-speed train. Asia and the Pacific islands, Africa, and South America and the Caribbean islands are presented by three different-sized squares that illustrate the size of their rain forests and the amount of deforestation in a decade. Forests cover 15 059 077 square miles, or 26 percent, of the land area; similar facts are given for deserts, pastures, ice caps, and other types of land areas; these facts are also depicted in an illustration that combines a number line and pictures. The border between the United States and Canada is longer than the Great Wall of China; in front of a drawing of the Great Wall is a line of 53 people holding hands with each person representing 10 000 people. The measure of all the world's coastlines is similar to the distance from Earth to the Moon; an illustration of this similarity also shows the Concorde and relates that it would take a week for the Concorde to fly the same distance. It should be noted that this description for "On the Surface" is for only *one* two-page spread. Some of the other twenty-two topics include disaster, big buildings, animal speed, growth and age, light and heavy, great and small, and population. Topics such as land and water speed and the solar system are presented in a foldout, four-page spread. This source is excellent for exploring number and size and for reexamining or expanding on how data can be represented. An index is included.

Burns, Marilyn. *The I Hate Mathematics! Book.* Illustrated by Martha Hairston. Boston: Little, Brown & Co., 1975.
See Number—Extensions and Connections: Number Theory, Patterns, and Relationships

————. *Math for Smarty Pants.* Illustrated by Martha Weston. Boston: Little, Brown & Co., 1982.
See Number—Extensions and Connections: Number Theory, Patterns, and Relationships

Clement, Rod. *Counting on Frank.* Milwaukee, Wis.: Gareth Stevens Children's Books, 1991.
See Number—Extensions and Connections: Number Concepts

Cushman, Jean. *Do You Wanna Bet? Your Chance to Find Out about Probability.* Illustrated by Martha Weston. New York: Clarion Books, 1991.
ISBN 0-395-56516-2 $14.95
★★ Single concept, 4–7

Whether flipping coins to decide what television program to watch or analyzing which events are "certain," "impossible," or "maybe," Danny and Brian become involved in everyday situations, both in and out of school, that involve probability. When the weather forecaster predicts a 60 percent chance that there will be enough snow to close school tomorrow, should Brian do his homework tonight? When it doesn't snow, was the forecaster wrong? These questions lead to collecting data to look at forecasting accuracy. Receiving invitations to two birthday parties to be held on the same day leads to exploring the likelihood that two students in a class will have the same birthday. An intercepted note being passed in class leads to a discussion of codes and the frequency of occurrence of certain letters in our language. The softball team's success this year is predicted by analyzing last year's batting averages. Cushman has woven a number of important probability concepts into an interesting story line with activities for the reader throughout the book. Drawings are used more to illustrate the story than to illustrate the concepts. A bibliography of resources that includes children's books and an index of concepts and activities are a welcome addition.

Fanelli, Sara. *My Map Book.* New York: HarperCollins Publishers, 1995.

ISBN 0-06-026455-1 $14.95
ISBN 0-06-026456-X   (library binding)   14.89

★★ Single concept, K–3

Fanelli's fantastic collages expand the definition of maps. Bright colors and unexpected discoveries can be found on each double-page spread. One map with arrows to indicate the four directions shows the layout of a medieval landscape, complete with a castle, a dragon, a treasure, and a car! Another colorful scale drawing, a "Map of My Bedroom," is complete with beds, toys, and potted plants. The "Map of My Family" provides a new twist as it shows a family tree and indicates the relationships among grandparents, parents, uncles, aunts, and cousins. A third interpretation is found in the "Map of My Day" where the page represents one whole day and is subdivided by parallel sections showing how the child spent each part of the day. Other "maps" include a color wheel, the regions of the heart, and a drawing of a dog, as well as more traditional maps such as my neighborhood and the seaside. This delightful book lends itself to some creative ways of representing spatial concepts and data and illustrating these relationships.

Griffiths, Rose. *Facts & Figures.* Illustrated by Peter Millard. Milwaukee: Gareth Stevens Publishing, 1994.

See Series and Other Resources: First Step Math

———. *Games.* Illustrated by Peter Millard. Milwaukee: Gareth Stevens Publishing, 1994.

See Series and Other Resources: First Step Math

Markle, Sandra. *Math Mini-Mysteries.* New York: Atheneum, 1993.

See Number—Extensions and Connections: Number Theory, Patterns, and Relationships.

———. *Measuring Up!: Experiments, Puzzles, and Games Exploring Measurement.* New York: Atheneum Books for Young Readers, 1995.

See Measurement: Measurement Concepts and Systems

Morgan, Rowland. *In the Next Three Seconds.* Illustrated by Rod Josey and Kira Josey. New York: Lodestar Books, 1997.

ISBN 0-525-67551-5 $13.99

★★★ Single concept, 3–8

"In the next three seconds 93 trees will be cut down to make the liners for disposable diapers.... In the next three minutes Americans will eat four and a half head of cattle as take-out hamburgers.... In the next three hours Americans will eat 600,000 lobsters...." Readers will be impressed as numerical wizard Morgan uses concepts of time in seconds, minutes, hours, days, ..., months, years, and decades in this intriguing collection of statistics and predictions. He forecasts life in three centuries and three thousand years and predicts that the space probe *Voyager 2* will be leaving our galaxy, the Milky Way, in the next three million years. Each double-page spread considers a different time interval. Most pages contain several predictions, with each statement enhanced by an illustration that reflects Rod Josey and Kira Josey's work with advertisers and packaging. Written in a style that is entertaining and humorous, this book also astonishes readers as it raises their level of concern about the preservation of our earth and its resources. Readers will be astounded not only by the book's predictions, which are research based, but also by the ease with which readers can develop their own predictions from given statistics. Morgan follows his introductory remarks with an example showing readers how easy it is to calculate their own unique and amazing predictions. He also suggests such resources for data as newspapers and includes some conversions for making comparisons, for example, a bathtub holds 160 quarts. *In the Next Three Seconds* can provide a delightful introduction to using real-life data to make predictions. This is a welcome edition for individuals familiar with Tom Parker's *In One Day,* which is now out of print.

Murphy, Stuart J. *The Best Vacation Ever.* Illustrated by Nadine Bernard Westcott. New York: HarperCollins Publishers, 1997.

See Series and Other Resources: MathStart

Pittman, Helena Clare. *Counting Jennie.* Minneapolis, Minn.: Carolrhoda Books, 1994.

ISBN 0-87614-745-7 $19.95

★★ Single concept, PS–3 (K–3)

From pies to pigeons and meatballs to monkeys, Jennie Jinks counted compulsively. Listening in to her thinking on the bus trip to school, we are made aware of the myriad of counting opportunities in the daily world of a bustling city. Jennie embraces them all! Only the bus driver's announcements of each stop offer rhythmic relief from Jennie's frenzy of counting. As Jennie reaches her destination, the school bell clangs; she runs but continues to count forty-three sidewalk squares, fourteen stairs, nine bulletin boards in the hallway.... As she opens the classroom door, the teacher smiles as he announces the day's attendance, "twenty-five ... counting Jennie." The richly colored watercolors skillfully reflect the intense activity of the morning rush hour and Jennie's need to classify and enumerate. The last three pages list all the sets that Jennie has counted. *Counting Jennie* provides a unique springboard to discuss and collect data in the world around us.

Scieszka, Jon. *Math Curse.* Illustrated by Lane Smith. New York: Viking, 1995.

See Number—Extensions and Connections: Number Concepts

Smoothey, Marion. *Graphs.* Illustrated by Ann Baum. New York: Marshall Cavendish, 1995.
See Series and Other Resources: Let's Investigate

————. *Statistics.* Illustrated by Ted Evans. New York: Marshall Cavendish, 1993.
See Series and Other Resources: Let's Investigate

Time-Life for Children Staff. *The Case of the Missing Zebra Stripes: Zoo Math.* Alexandria, Va.: Time-Life for Children, 1992.
See Series and Other Resources: I Love Math

————. *From Head to Toe: Body Math.* Alexandria, Va.: Time-Life for Children, 1993.
See Series and Other Resources: I Love Math

————. *How Do Octopi Eat Pizza Pie: Pizza Math.* Alexandria, Va.: Time-Life for Children, 1993.
See Series and Other Resources: I Love Math

————. *The Mystery of the Sunken Treasury: Sea Math.* Alexandria, Va.: Time-Life for Children, 1993.
See Series and Other Resources: I Love Math

————. *Play Ball: Sports Math.* Alexandria, Va.: Time-Life for Children, 1993.
See Series and Other Resources: I Love Math

————. *Right in Your Own Backyard: Nature Math.* Alexandria, Va.: Time-Life for Children, 1993.
See Series and Other Resources: I Love Math

Vorderman, Carol. *How Math Works.* Pleasantville, N.Y.: Reader's Digest Association, 1996.
See Series and Other Resources: Resources

Walpole, Brenda. *Speed.* Illustrated by Dennis Tinkler and Chris Fairclough. Milwaukee, Wis.: Gareth Stevens, 1995.
See Series and Other Resources: Measure Up with Science

————. *Temperature.* Illustrated by Dennis Tinkler and Chris Fairclough. Milwaukee, Wis.: Gareth Stevens, 1995.
See Series and Other Resources: Measure Up with Science

Number Theory, Patterns, and Relationships

Patterns, patterns, and more patterns are the common thread in this category. This section is dominated by quality books that explore number patterns and relationships. These patterns lead to topics in number theory and to ideas involving functions. The simplicity with which these ideas are presented makes these concepts accessible and appealing to children and adults. Sequences are the focus of *Anno's Magic Seeds*. Calmenson's *Dinner at the Panda Palace* includes the sum of a sequence, the sum of the first ten counting numbers. Number patterns and relationships for young children are found in *Pattern* by Pluckrose; more sophisticated ones can be explored in Anno's books. Both Anno and the Time-Life for Children Staff include function machines; other investigations and information involving functions and algebra are presented by Smoothey and Adler.

Geometric patterns involving doubling are embedded in the intriguing folktales by Barry, Birch, and Demi. Other connections between number and geometry are implied such as star polygons in Carle's *Draw Me a Star* and square and rectangular numbers in books by Hulme, Mathews, and Pinczes. Representations and patterns are used by the Annos in *Anno's Mysterious Multiplying Jar* to explore factorials. Ratios, a relationship between numbers, are dramatically illustrated by Norden and Ruschak and Wells. Applications of these ideas are included in the works by Smoothey and Walpole. Readers should be captivated by the many problem-solving situations involving number theory, patterns, and other relationships posed by Burns in *The I Hate Mathematics! Book* and *Math for Smarty Pants*, Hayes and Hayes in Number Mysteries, and others.

The following books are out of print but are considered exceptional:

Charosh, Mannis. *Number Ideas through Pictures*
Frédérique and Papy. *Graph Games*
Froman, Robert. *A Game of Functions*

Lewin, Betsy. *Cat Count*

Srivastava, Jane Jonas. *Number Families*

Weiss, Malcolm E. *666 Jellybeans! All That? An Introduction to Algebra*

Copies may be available in your library. See the first edition of *The Wonderful World of Mathematics* for annotations.

Adler, Irving. *Mathematics.* Illustrated by Ron Miller. New York: Doubleday, 1990.

ISBN 0-385-26142-X	$12.95
ISBN 0-385-26143-8 (library binding)	12.99

★★ Multiconcept, (4–8)

Connections between number and geometry with extensions to simple computer programs are cleverly presented in this book on mathematics. Concise text is set off in pastel rectangular areas that are imposed over the attractive illustrations. Historical uses of number and space are briefly described in the opening paragraph against a background collage picturing large numerals, an abacus, the ruins of Stonehenge, a scene of a shepherd and his flock, a surveyor and the Egyptian pyramids, and an ancient manuscript. Another collage illustrating the golden ratio includes Piet Mondrian's *Composition in White, Black and Red,* the Parthenon, a regular pentagon and its diagonals, and a line divided so that the ratio of its two parts is the same as that of the longer part to the whole. Natural numbers are compared to a ladder that never ends; the narrative clearly describes how to write a calculator code and a BASIC program for counting from one. Background illustrations include a ladder that reaches to the stars and beyond with the numbered steps gradually fading in the distance to imply infinity and a flowchart that diagrams the decision processes of the computer program. Even and odd numbers are introduced and represented by counters; modifications to the natural-number computer program are made so that the sequence of even or odd numbers can be printed. Later, square numbers are related to the sequence of odd numbers and the program is modified to add the sequence of odd numbers. This sequence is later connected to triangular numbers. The transitions from topic to topic are smooth and provide good connections between number and geometry as well as connections among number topics. Other topics that are explored include integers; rectangular numbers; primes; perfect, amicable, and sociable numbers; Fibonacci numbers; and the number of degrees in a polygon. Adler provides an excellent overview of various topics in mathematics that is appropriate for the novice or the expert. An index is included.

Anno, Masaichiro, and Mitsumasa Anno. *Anno's Mysterious Multiplying Jar.* Illustrated by Mitsumasa Anno. New York: Philomel Books, 1983.

ISBN 0-399-20951-4 $16.95

★★★ Single concept, 3+

A story of imagination and mathematics is beautifully illustrated and poetically stated. A mysterious jar is the vehicle that transports the reader into the world of factorials. The jar contains rippling water that becomes a wide, deep sea with one island containing two countries. "Within each country there were 3 mountains.... On each mountain there were 4 walled kingdoms." The pattern continues and accumulates until each of nine boxes contains ten jars. Through the text and the illustrations, the reader becomes aware that the pattern is rapidly expanding. These ideas are then extended by using arrays and factorial notation to explore the relative size of each of these groups. To show that there were four walled kingdoms on each of six, or 3!, mountains, the array is four times larger than the previous array. 4! = 4 × 3! = 4 × 3 × 2 × 1 = 24. In the afterword, the authors introduce additional ideas to explore with this multiplication pattern.

Anno, Mitsumasa. *Anno's Magic Seeds.* New York: Philomel Books, 1994.

ISBN 0-399-22538-2 $15.95

★★★ Single concept, PS–2 (2–4)

Some of the most exceptional children's book about mathematics have been written by Anno; *Anno's Magic Seeds* adds to this collection—it is an excellent book on number patterns. The pattern begins as a wizard gives Jack a magic seed, which he plants. It blooms and produces 2 seeds. Jack eats 1 seed and plants the other; this pattern continues for five years. After thinking about the situation, Jack plants both seeds and continues to harvest 2 seeds for each 1 planted; he eats 1 and plants the rest. As this pattern is continued, the reader is asked to determine how many seeds will be planted and how many will be harvested. The pattern changes again when Jack and Alice are married; 2 seeds are given to each wedding guest, 2 seeds are now eaten each year, and a storehouse is built for the reserved seed. Questions continue to be posed; for each season, new information is given about the number of seeds stored and the number eaten. The calculated solution can be verified by counting the seeds in the watercolor illustrations. The text and one scene show 51 seeds in the storehouse (five bags of 10 seeds plus 1 more seed), ten bags of 10 seeds being taken to the market, 3 seeds to be eaten (1 seed each for Jack, Alice, and the baby), and 120 seeds for next year's planting (twelve sections of a drying rack with each section holding five plants each of which produced 2 seeds). The family continues to prosper until a hurricane wipes out the crop and all but 10 seeds that Jack rescued from the storehouse. With these the family starts over. In the author's notes, Anno writes that the story mirrors the evolution of commerce and trade in civilization; the magic power of the seeds to reproduce in quantity is well illustrated.

————. *Anno's Math Games*. New York: Philomel Books, 1987.
ISBN 0-399-21151-9 $19.95
★★★ Multiconcept, PS–3 (1–5)

The concepts in this 104-page book are beautifully illustrated, cleverly presented, and interrelated to ideas in the world around us. There is a section to help parents, teachers, or other older readers discuss and extend the ideas in this book. The first section, "What is Different?" contains sets from which the reader is to determine which object is different and how it is different. Many ideas and levels of sophistication are included; for example, an animal may not belong to the set because it flies or is extinct. Folklore and inventions introduce the second section, "Putting Together and Taking Apart." Charts record all twenty-five examples of the possibilities of crossing five colors with five articles of clothing (Cartesian products). The reader can extend "Putting Together" by exploring how five puzzle pieces can fit into various shapes. Playing cards introduce sequences and cardinal and ordinal numbers in "Numbers in Order." These concepts are extended in examples involving sequences and locations, such as grids, a theater, and apartment houses. The last section, "Who's the Tallest?" introduces attributes that we can measure, such as weight and time, and discusses concepts that cannot be measured, such as scary or sad. A direct comparison of heights leads to indirect measures that use bar graphs, spring scales, and other measuring instruments. One of the last examples, on proportion, compares the relative numbers of sugar cubes in different-sized containers. A wealth of ideas appears in this book that can be extended to a number of situations; for this reason, it may be desirable to explore one chapter at a time. Readers of many ages should enjoy the challenges.

————. *Anno's Math Games II*. New York: Philomel Books, 1989.
ISBN 0-399-21615-4 $19.95
★★★ Multiconcept, 1–4

Kriss and Kross, the two friends who guided readers through the explorations in *Anno's Math Games,* continue their investigations. This 104-page, five-chapter sequel is beautifully illustrated, cleverly presented, and similar in format and style to the first volume. The section to help parents, teachers, or other older readers discuss and extend the ideas in this book is very interesting reading. The first chapter is about relationships; a "magic machine" defines relationships of age, number, doubling, shape, or "one more than" between two sets. There is some exploration of inverse relationships. "Compare and Find Out" emphasizes similarities and differences. On each pair of opposing pages, the reader makes comparisons of quantity, size, arrangement, type, position, and elements. The pages offer attractive and interesting challenges. The third chapter, "Dots, Dots, and More Dots," looks at points as a basic element of design— from the dots of the impressionist Seurat to the atoms composing all matter. One-to-one correspondence, rational counting, and place value are uniquely explored in "Counting with Circles." A different way to count such substances as sugar, salt, or water is explored in "Counting Water." Whether comparing congruent containers or counting small containers, creating calibrated measuring instruments or using different-sized measures, Anno's designs for exploring measures of capacity are a delight.

Barry, David. *The Rajah's Rice: A Mathematical Folktale from India.* Illustrated by Donna Perrone. New York: Scientific American Books for Young Readers, 1994.

ISBN 0-7167-6568-3 $15.95

★★ Single concept, 2–4 (2–6)

Perrone's rich illustrations reflecting the Rajah's court and the Indian environment enhance Barry's excellent adaptation of this classic folktale. The power of doubling is explored through the familiar checkerboard problem, which is embedded in the story of Chandra, the official bather of the Rajah's elephants. When Chandra saves the elephants from an infection undiagnosed by the medical men, the Rajah throws a celebration and asks her to name her reward. Chandra, who has been troubled by the poverty of her countrymen, notices a checkerboard and asks for the amount of rice resulting from placing 2 grains of rice on the first square, 4 on the second square, 8 on the third square, and so on. The illustrations reflect the increasing amount of rice with 256 grains covering the last square in the first row. Rather than count rice grains to fill the second row, Chandra suggests that 256 grains are one teaspoon, and the Rajah commands that the rice be measured in teaspoons instead. By the end of the second row, a bowl is used to measure the rice, and by the third row, a large wheelbarrow is used. Aware of his error, the Rajah asks to be released from his original promise because the storehouse has been emptied by the middle of the fifth row. A final page shows equivalent measures that could be used for each row; the 256 bowls fill 1 wheelbarrow, 256 wheelbarrows fill 1 festival hall, 256 halls fill 1 palace, and so on, to 256 Manhattan Islands fill 1 Mount Kilimanjaro. These benchmarks help illustrate the magnitude of the measures and the power of doubling.

Birch, David. *The King's Chessboard.* Illustrated by Devis Grebu. New York: Dial Books for Young Readers, 1988.

ISBN 0-8037-0367-8   (library binding) $10.95
ISBN 0-14-054880-7   (paper, Puffin)   4.99

★★ Single concept, PS–2 (2–6)

An ancient land, now called India, is the setting for the story of a king who ponders a gift for his wise man in spite of the man's objections. At the king's insistence, the wise man acquiesces and asks for one grain of rice for the first square of the king's chessboard on the first day, two grains for the second square on the next day, four grains for the third square on the third day, eight grains for the fourth square on the fourth day, and so on, doubling the amount of rice each day for each square on the chessboard. The king is perplexed about the proposed amount of rice for sixty-four days, but he is too proud to ask. In the days that follow, the Royal Weigher changes to measuring the rice by weight instead of counting the grains as he realizes the magnitude of the increasing numbers. On consultation with the Grand Superintendent and the Chief Mathematician, the king finally realizes that his promise cannot be kept. The graceful illustrations reflect this classic problem as the tale unfolds.

Burns, Marilyn. *The I Hate Mathematics! Book.* Illustrated by Martha Hairston. Boston: Little, Brown & Co., 1975.

ISBN 0-316-11740-4 $17.95
ISBN 0-316-11741-2   10.95

★★★ Multiconcept, 5–8

"Street Math" and "Things to Do When You Have the Flu" are two of the chapters in this challenging book that sets out to show the reader how much fun mathematics can be. Estimation and prediction are used in experiments on how close a person can approach pigeons or strangers on a street. This chapter on street math also includes patterns, definitions, estimation with volume and number, data collection, topology, combinations, and permutations. Each page presents one or more intriguing facts, questions, or experiments involving ratios, volume, estimation and measurement, spatial relationships, clock arithmetic, strategy games, and probability. The text is upbeat, and drawings add humor and clarity throughout the 127-page book. This should be a sure remedy to "I hate math" attitudes.

———. *Math for Smarty Pants.* Illustrated by Martha Weston. Boston: Little, Brown & Co., 1982.

ISBN 0-316-11738-2 $17.95
ISBN 0-316-11739-0   (paper)   11.95

★★★ Multiconcept, 5–8

"Arithmetic with a Twist," the first chapter in this absorbing book, explores the topics of looping, one-dollar words, and chummy numbers along with intriguing stories about three little pig eyes and incredible calculators. Geometry is the theme of "The Shapes of Math," as problems are posed about shapes, properties, estimation, and triangular and square numbers. Graphic explanations of factorials and offbeat facts about pencils and populations are incorporated into a chapter on statistics and probability. Other chapters focus on strategy games or logic puzzles, which include paradoxes and number theory. The puzzles in "Math Trickery" should fascinate readers and challenge them to consider, "Why?" A final chapter includes benchmarks for developing number sense about large numbers. Drawings are used throughout to provide humor, explanation, and appeal. This fascinating collection of problems and facts should captivate many middle school students.

Calmenson, Stephanie. *Dinner at the Panda Palace.* Illustrated by Nadine Bernard Westcott. New York: Scholastic, 1991.

See Early Number Concepts: Counting—Operations

Carle, Eric. *Draw Me a Star.* New York: Philomel Books, 1992.

See Geometry and Spatial Sense: Shapes

Demi. *One Grain of Rice: A Mathematical Folktale.* New York: Scholastic Press, 1997.

ISBN 0-590-93998-X $19.95

★★ Single concept, K–3 (2–6)

Exquisite artwork is the centerpiece of Demi's brilliant adaptation of the traditional Indian tale on doubling. In this version, year after year, the raja takes the farmers' rice, leaving them with enough only to eat. When a famine arrives, he does not share the accumulated rice in the royal storehouses. As a reward for an honest deed, Rani, a village girl, asks the raja for 1 grain of rice for the first day, 2 grains for the second day, 4 grains for the third day, and so on, for thirty days; the raja agrees to the seemingly modest request. Stunning illustrations show the rice being delivered by native animals like a parrot, a peacock, a heron, a leopard, a tiger, a lion, and a goat with a cart. The size and number of animals needed for the delivery increase with each passing day. Many references are given to the number of grains of rice as well as to the size of the quantity; on the sixteenth day, 32 768 grains, or two full bags, of rice were delivered; on the twenty-fourth day, 8 388 608 grains in eight baskets were carried by eight royal deer. Soon, 536 870 912 grains are carried on 256 elephants; the elephants are arranged in twenty-five columns of 10 across four fold-out pages. A final page shows the number of grains for each of the thirty days and the answer for the accumulated sum. Rich colors with reds and golds dominate; Demi's inspiration for the artwork was Indian miniature paintings from the sixteenth and seventeenth centuries. Each illustration is framed with a gold border and a second red border; in some illustrations the figures overlap or are placed outside the borders. On one fold-out page, the combination of color and design makes the forty trios of camels carrying baskets on the twenty-ninth day appear three-dimensional.

Griffiths, Rose. *Number Puzzles.* Illustrated by Peter Millard.
Milwaukee, Wis.: Gareth Stevens Publishing, 1995.

See Series and Other Resources: First Step Math

Hayes, Cyril, and **Dympna Hayes**. *Number Mysteries.* Illustrated by
Peggy McEwen, Shane Doyle, Rick Rowden, and Jodi Shuster. Niagara
Falls, N.Y.: Durkin Hayes Publishing, 1988.

ISBN 0-88625-145-1   (paper) $2.95

★★ Multiconcept, 4–6

Challenging problems for children appear in the text and lively
illustrations. Unique and motivating topics include number puzzles
involving number and operation sense, beginning probability, and
ratio. Spatial relationships are embedded in some of the problems. To
solve these problems, the readers will soon find themselves using
such strategies as number lines, diagrams, models, graphs, tables,
and patterns. The illustrations are colorful and humorous. However,
since each page presents several problems to be solved, sometimes
the illustrations appear a bit cluttered. Solutions to some problems
are included at the end of the book.

Heath, Royal Vale. *MatheMagic: Magic, Puzzles, and Games with
Numbers.* New York: Dover Publications, 1953.

ISBN 0-486-20110-4 $3.95

☆ Multiconcept, 2+

Number theory is the basis of many problems in this 126-page
collection of number puzzles. Place value, base two, properties of
operations, doubling, remainders, and palindromes are some of the
other embedded mathematics concepts. Mental computation is
included throughout the book, and one section describes "easy" ways
to solve problems mentally. A section called "Number Symphonies"
simply shows patterns involving primarily the multiplication of very
large numbers. Magic is the theme of one section of ten number
tricks to be played with another individual. Directions allow the
reader to appear to be able to read the other person's mind. A
numerical explanation is given of how the tricks work; more
advanced readers could use algebra to determine why they work.
Thirty-seven pages explore a variety of magic squares and numbers
in other patterns. Although parts of the book could entertain second
graders, it is not designed for young children. The examples are
multistep problems and require fairly sophisticated procedures,
including square roots. This interesting resource for number tricks
could motivate middle school children. Illustrations are minimal, but
some black-and-white drawings are used for diagrams and interest.

Hong, Lily Toy. *Two of Everything: A Chinese Folktale.* Morton Grove, Ill.: Albert Whitman & Co., 1993.
See Number—Extensions and Connections: Operations

Hulme, Joy N. *Sea Squares.* Illustrated by Carol Schwartz. New York: Hyperion Books for Children, 1991.
See Number—Extensions and Connections: Operations

Kallen, Stuart A. *Mathmagical Fun.* Edina, Minn.: Abdo & Daughters, 1992.
ISBN 1-56239-129-1 $12.94
☆ Single concept, 4–7

Part of the puzzle series titled Giggles, Gags, and Groaners, this 32-page book focuses on the magic of numbers. Thirteen "magical math tricks" involving number are carefully described. Some of the tricks can be readily explained through patterns and observations. For example, "numbers on the table" involves sequentially writing the numerals from 1 through 9 in a 3-by-3 pattern on a sheet of paper. When the paper is torn into nine pieces and placed face down on the table, the even or odd numbers can be readily identified by the number of ragged edges on each piece. The tricks appear to involve mind reading but when analyzed are seen to involve place value, factors, or other number theory ideas. For "down the line" one person selects any three-digit number and uses the calculator to multiply it by 7; the calculator is handed to the next person who multiplies the result by 11, and a third person multiplies by 13. The magician can read the last product to identify the original three-digit number. All these "tricks" can (and should) be analyzed and explained mathematically to show that mathematics is not a mystery.

Markle, Sandra. *Math Mini-Mysteries.* New York: Atheneum, 1993.
ISBN 0-689-31700-X $14.95
★★ Multiconcept, 3–7

Natural science is the basis for many of Markle's "mysteries," which are defined as problems waiting for a solution. Readers are challenged with this hands-on investigative approach to mathematics. An opening page lists problem-solving strategies that readers are encouraged to copy and use as they read the subsequent problems and work through the necessary stages to a solution. Many of the problems are embellished with interesting facts and explanations about history, science, and other connections to our world. Yellowstone National Park is the basis for two problems: using an expression to determine when Old Faithful will erupt again and reading a map to determine travel times for a grand tour. Ratios are explored with measures from the Statue of Liberty and the reader. Arithmetic skills are essential for seeking solutions for magic squares or determining the costs of feeding a zoo's animals. Directions are given for constructing mazes. Geometry, ratio, measurement, patterns, data collection, and interpretation are some of mathematics topics explored. Black-and-white photographs, charts, graphs, and maps complement the text appropriately. Directions for each problem and subsequent activity are clearly and simply stated; one confusing exception is the diagrams and directions for cutting paper snowflakes to explore radial symmetry. Many problems are presented; the print is small. The total effect of this amount of reading may deter all but the avid mathematician. It would be helpful to view *Math Mini-Mysteries* as a collection of problems from which readers can pick and choose the ones that they wish to pursue. A table of contents and an index are included; answers can be found throughout the text.

Mathews, Louise. *Bunches and Bunches of Bunnies.* Illustrated by Jeni Bassett. New York: Scholastic, 1978.

See Number—Extensions and Connections: Operations

McKibbon, Hugh William. *The Token Gift.* Illustrated by Scott Cameron. New York: Annick Press, 1996.

ISBN 1-55037-499-0 $16.95
ISBN 1-55037-498-2   (paper)   6.95

★★ Single concept, 1–6

McKibbon creates another version of the classic story of the doubling of the grains of rice on a chessboard. The story is set in ancient India, where a young man, Mohan, invents a board game that involves strategic thinking. He calls the game "Chaturanga" and gradually the rules of the game are refined and the popularity of the game spreads. When the king wishes to honor the game's inventor, Mohan, now a wealthy old man, refuses. But on the king's insistence of a gift, Mohan requests a grain of rice for the first square of the chessboard, two grains for the second square, four grains for the third square, and so on, for all sixty-four squares. After the sixteenth square, with 32 768 grains, a bag of grain is suggested as a substitute for the individual grains of rice until there are 32 768 bags at the thirty-second square. For the thirty-third through the forty-eighth squares, shiploads of rice are discussed as substitutes. And finally the royal farm advisor observes that it would take half a year for the entire kingdom to raise that much rice. Consequently, for the rest of the squares the royal mathematician uses doubling to determine the number of years it will take to raise the necessary amount of rice. All this information is clearly shown in a four-column chart in one illustration; at the end of each column are the accumulating sums. The predicament is eventually resolved to everyone's satisfaction. The afterword includes additional information about this legend involving the game of chess and the reward for its inventor. The ancient culture and the story line are reflected in the rich brown tones of the dark, majestic illustrations, which were originally done in oil.

Norden, Beth B., and **Lynette Ruschak.** *Magnification: A Pop-Up Lift-the-Flap Book.* Illustrated by Wendy Smith-Griswold, Matthew Bens, and Robert Hynes. New York: Lodestar Books, 1993.
See Measurement: Measurement Concepts and Systems

Pinczes, Elinor J. *A Remainder of One.* Illustrated by Bonnie MacKain. Boston: Houghton Mifflin Co., 1995.
See Number— Extensions and Connections: Operations

————. *One Hundred Hungry Ants.* Illustrated by Bonnie MacKain. Boston: Houghton Mifflin Co., 1993.
See Number—Extensions and Connections: Operations

Pluckrose, Henry. *Pattern.* Illustrated by Chris Fairclough. Chicago: Children's Press, 1995.
See Series and Other Resources: Math Counts

Sharp, Richard M., and **Seymour Metzner.** *The Sneaky Square and 113 Other Math Activities for Kids.* Blue Ridge Summit, Pa.: TAB Books, 1990.

ISBN 0-8306-8474-3	$15.95
ISBN 0-8306-3474-6 (paper)	8.95

☆ Multiconcept, 3–7

"Traps and Conundrums," "Number Problems," "Geometricks," "Combination Puzzles," "Positioning," and "Mathematical Relationships" are the chapter titles in this book of problems and puzzles that includes many classic mathematical problems. A number of the problems appear to be tricks, but on closer examination, the explanations for their solutions are embedded in algebra, spatial reasoning, logical reasoning, number theory, and number relationships. Each problem shows the difficulty level (low, medium, or high) and the materials needed (usually paper and pencil); an example or diagram often illustrates the situation. A brief explanation helps answer why the problem works or relates it to a specific mathematical topic. Black-and-white drawings throughout the book illustrate the problem or add interest. Processes of problem solving, mental computation, and mathematical reasoning will be needed to solve this challenging collection.

Smoothey, Marion. *Codes and Sequences.* Illustrated by Ann Baum. New York: Marshall Cavendish, 1995.
See Series and Other Resources: Let's Investigate

————. *Estimating.* Illustrated by Ann Baum. New York: Marshall Cavendish, 1995.
See Series and Other Resources: Let's Investigate

———. *Graphs.* Illustrated by Ann Baum. New York: Marshall Cavendish, 1995.
See Series and Other Resources: Let's Investigate

———. *Maps and Scale Drawings.* Illustrated by Ann Baum. New York: Marshall Cavendish, 1995.
See Series and Other Resources: Let's Investigate

———. *Number Patterns.* Illustrated by Ted Evans. New York: Marshall Cavendish, 1993.
See Series and Other Resources: Let's Investigate

———. *Numbers.* Illustrated by Ted Evans. New York: Marshall Cavendish, 1993.
See Series and Other Resources: Let's Investigate

———. *Ratio and Proportion.* Illustrated by Ann Baum. New York: Marshall Cavendish, 1995.
See Series and Other Resources: Let's Investigate

———. *Time, Distance, and Speed.* Illustrated by Ted Evans. New York: Marshall Cavendish, 1993.
See Series and Other Resources: Let's Investigate

Stienecker, David L. *Multiplication.* Illustrated by Richard Maccabe. New York: Benchmark Books, 1996.
See Series and Other Resources: Discovering Math

———. *Numbers.* Illustrated by Richard Maccabe. New York: Benchmark Books, 1996.
See Series and Other Resources: Discovering Math

Time-Life for Children Staff. *The House That Math Built: House Math.* Alexandria, Va.: Time-Life for Children, 1993.
See Series and Other Resources: I Love Math

———. *Look Both Ways: City Math.* Alexandria, Va.: Time-Life for Children, 1992.
See Series and Other Resources: I Love Math

———. *The Mystery of the Sunken Treasure: Sea Math.* Alexandria, Va.: Time-Life for Children, 1993.
See Series and Other Resources: I Love Math

———. *Pterodactyl Tunnel: Amusement Park Math.* Alexandria, Va.: Time-Life for Children, 1993.
See Series and Other Resources: I Love Math

————. *The Search for the Mystery Planet: Space Math.* Alexandria, Va.: Time-Life for Children, 1993.
See Series and Other Resources: I Love Math

————. *See You Later Escalator: Mall Math.* Alexandria, Va.: Time-Life for Children, 1993.
See Series and Other Resources: I Love Math

Vorderman, Carol. *I Love Math.* Pleasantville, N.Y.: Reader's Digest Association, 1996.
See Series and Other Resources: Resources

Wahl, John, and Stacey Wahl. *I Can Count the Petals of a Flower.* Reston, Va.: National Council of Teachers of Mathematics, 1985.
See Early Number Concepts: Counting—Other Number Concepts

Walpole, Brenda. *Counting.* Illustrated by Dennis Tinkler and Chris Fairclough. Milwaukee: Gareth Stevens, 1995.
See Series and Other Resources: Measure Up with Science

————. *Distance.* Illustrated by Dennis Tinkler and Chris Fairclough. Milwaukee: Gareth Stevens, 1995.
See Series and Other Resources: Measure Up with Science

Wells, Robert E. *Is a Blue Whale the Biggest Thing There Is?* Morton Grove, Ill.: Albert Whitman & Co., 1993.
See Measurement: Measurement Concepts and Systems

————. *What's Smaller than a Pygmy Shrew?* Morton Grove, Ill.: Albert Whitman & Co., 1995.
See Measurement: Measurement Concepts and Systems

White, Laurence B., and **Ray Broekel.** *Math-a-Magic: Number Tricks for Magicians.* Illustrated by Meyer Seltzer. Morton Grove, Ill.: Albert Whitman & Co., 1990.
See Number—Extensions and Connections: Reasoning—Classification, Logic, and Strategy Games

Reasoning—
Classification,
Logic, and
Strategy Games

During the Middle Grades
Graduate Program, a time
period was set aside to browse over
new replacement units, laser disks, and other resources. One of the teachers,
Tina, started reading children's books by Anno and then borrowed *Anno's
Hat Tricks*. The last logic problem was challenging, so later that afternoon
she discussed it with the teachers in her carpool. The next morning the
discussion continued during their morning drive, and the first twenty
minutes in their problem-solving course was rescheduled so that the entire
class could enjoy the book and work on the problem. Needless to say, Tina
introduced *Anno's Hat Tricks* to her middle-grades students.

To entertain Jermone, the teenaged son of overnight guests, I handed him
Talbot's *It's for You: An Amazing Picture-Puzzle Book*. "Thank you" was the
last comment heard from him for the next three hours. Later, his mother
became intrigued with the book, and well after midnight mother and son
were still discussing possible solutions. The next morning they were found
poring over the book again.

Quality children's books are captivating for individuals of *all* ages.
Well-designed mathematical tasks coupled with complementary
illustrations are two of the main reasons for their appeal. Hoban's *Is It
Red? Is It Yellow? Is It Blue?* and *Is It Rough? Is It Smooth?* and
Pluckrose's *Sorting* introduce classification and sorting, beginning ideas in
mathematics for children. Classification and Venn diagrams are used by
the Time-Life for Children staff to solve problems involving logical
reasoning. In *Math-a-Magic: Number Tricks for Magicians*, White and
Broekel present strategy games with a flair. Books containing strategy
games or puzzle problems that involve reasoning are also included in this
category.

The following books are out of print but are considered exceptional:

Charosh, Mannis. *Mathematical Games for One and Two*
Froman, Robert. *Venn Diagrams*
Gersting, Judith L., and Joseph E. Kuczkowski. *Yes-No, Stop-Go: Some Patterns in Mathematical Logic*

Copies may be available in your library. See the first edition of *The Wonderful World of Mathematics* for annotations

The Book of Classic Board Games. Palo Alto, Calif.: Klutz Press, 1991.
ISBN 0-932592-94-5 $15.95
★★★ Single concept, K+

Playing pieces and a container are attached to the sturdy cardboard cover; the fifteen game boards are printed on good-quality paper. The book is spiral bound, so it lies conveniently flat. This thoughtful design and format are typical of Klutz Press. Also typical is the humor throughout the text and illustrations. Each artfully designed game board is preceded by a set of simple directions that include the number of players, how to start, and how to play, including sample plays. The table of contents suggests age levels and notes the level of difficulty of each game. The criteria for including games in this book were that the games require strategy and that they are multilayered for players of different experiences. The selected games have had universal appeal over a long time; a brief history and the current status of each game are included. It appears that the staff also met the goals of providing games that are fun and involve logical thinking.

Bulloch, Ivan. *Games.* New York: Thomson Learning, 1994.
See Series and Other Resources: Action Math

Burns, Marilyn. *Math for Smarty Pants.* Illustrated by Martha Weston. Boston: Little, Brown & Co., 1982.
See Number—Extensions and Connections: Number Theory, Patterns, and Relationships

Carle, Eric. *My Very First Book of Colors.* New York: HarperCollins Children's Books, 1985.
ISBN 0-694-00011-6 $4.95
☆ Single concept, PS–K

Carle develops this beginning classification book in the same format as the other books in this My Very First Book of ... series. The cardboard pages are cut horizontally and have rectangles of different colors on the top half of the pages. All these half pages are one color except one page showing nine swatches of different colors. Vibrant pictures of different objects are pictured on the bottom half of the pages. The child can turn the half pages separately to match the objects with the appropriate colors. Most illustrations are predominantly one color, but often more than one correct match is possible.

Griffiths, Rose. *Games.* Illustrated by Peter Millard. Milwaukee: Gareth Stevens Publishing, 1994.

See Series and Other Resources: First Step Math.

Hayes, Cyril, and Dympna Hayes. *Number Mysteries.* Illustrated by Peggy McEwen, Shane Doyle, Rick Rowden, and Jodi Shuster. Niagara Falls, N.Y.: Durkin Hayes Publishers, 1988.

See Number—Extensions and Connections: Number Theory, Patterns, and Relationships

Hoban, Tana. *Is It Red? Is It Yellow? Is It Blue?* New York: Greenwillow Books, 1978.

ISBN 0-688-80171-4	$13.95
ISBN 0-688-84171-6 (library binding)	13.88

☆ Single concept, K–3

Red, yellow, blue, orange, green, and purple are the vibrant colors captured in Hoban's photographs of familiar objects. The colors are named only on an opening page in this wordless text. Readers are to find objects in each scene that match each color shown in the lower border of each page. The appealing photographs offer an opportunity for readers to classify objects by color, which could be readily extended to classifying further by some other attribute.

_____. *Is It Rough? Is It Smooth? Is It Shiny?* New York: Greenwillow Books, 1984.

ISBN 0-688-03823-9	$10.25
ISBN 0-688-03824-7 (library binding)	10.88

☆ Single concept, PS–1

Bright pennies, rugged elephant skin, weathered tree bark, and sticky caramel apples are some of the objects shown in Hoban's photographic essay on texture. Strong images clearly convey the attributes of each surface. This no-text picture book could be used to encourage classification activities on texture or extended to the exploration of curved and flat surfaces.

_____. *Of Colors and Things.* New York: Greenwillow Books, 1989.

ISBN 0-688-07534-7	$15.00
ISBN 0-688-07535-5 (library binding)	14.93

★★ Single concept, PS–1

Photographed against a white background, each object is sharply defined in a color close-up. On each page, four photographs are separated by wide perpendicular bands of color; objects on any one page are predominantly one color. The brilliant photographs of grays, reds, blues, oranges, blacks, yellows, greens, and browns should invite comparisons among the objects on each page or on other pages. Since the objects are familiar, readers should be motivated to find other objects in their environment that have similar attributes.

Kallen, Stuart A. *Mathmagical Fun.* Edina, Minn.: Abdo & Daughters, 1992.

See Number—Extensions and Connections: Number Theory, Patterns, and Relationships

Nozaki, Akihiro. *Anno's Hat Tricks.* Illustrated by Mitsumasa Anno. New York: Philomel Books, 1985.

ISBN 0-399-21212-4 $15.95

★★★ Single concept, 3+

Three characters interact in this unique book—Tom, Hannah, and Shadowchild, the reader. "What color is your hat?" is the question posed to Shadowchild in each activity in the book, which introduces binary logic. One or both of the other two characters, wearing either red or white hats, are pictured on each layout. The "shadow" of the reader and his or her hat stretches across each illustration. Since shadows do not show color, the color of Shadowchild's hat cannot be determined by looking at the illustration. Through a combination of illustrations and text, the reader can discern how many red and white hats there are and how many are already accounted for. By analyzing the various clues, Shadowchild can determine the color of his or her own hat. The questions gradually become more complex as the book progresses but always involve two choices—red hats or white hats. The author includes notes to parents and other older readers that extend the concepts and use tree diagrams to discuss some of the examples. The clever, clearly written text with its simple illustrations should motivate readers to explore the intriguing problems posed. This is another excellent example of significant mathematical ideas presented in a style and format suitable for children.

Pluckrose, Henry. *Sorting.* Illustrated by Simon Roulstone. Chicago: Children's Press, 1995.

See Series and Other Resources: Math Counts

Reid, Margarette S. *The Button Box.* Illustrated by Sarah Chamberlain. New York: Dutton Children's Books, 1990.

ISBN 0-525-44590-0 $13.99
ISBN 0-14-055495-5 (paper, Puffin Books) 4.99

★★ Single concept, PS–3

A young boy's fascination with the contents of Grandma's button box is the basis of this classification story. Colorful illustrations and simple text invite the reader to sort, classify, sequence, and count buttons. Many ways to classify buttons are suggested: size, texture, number of holes, use, and composition. One illustration shows an interesting pattern of large to small buttons symmetrically arranged in rows of one, three, five, seven.... Reid provides an opportunity that can be used as a springboard to other classification activities.

Sis, Peter. *Beach Ball.* New York: Greenwillow Books, 1990.

ISBN 0-688-09181-4 $12.95
ISBN 0-688-09182-2   (library binding)   12.88

☆ Multiconcept, K–3

As Mary runs along the sandy shoreline in pursuit of her blown-away beach ball, she passes through a series of scenes filled with challenges to the reader. The pastel illustrations are filled with all types of interesting characters and humorous situations. In one busy scene, opportunities to count objects from one through ten include a school of nine fish swimming along the shore, a lady with eight pairs of sunglasses, the lifeguard with seven life preservers, and a man under two sun umbrellas. In another two-page spread, prisms, spheres, cylinders, tori, and other shapes are part of the seascape focusing on shape. Triangles, hearts, rectangles, hexagons, circles, and sectors of circles are included in the design and in the shapes of blankets, sails, and other objects. Another page shows some concepts of opposites that include high and low, long and short, round and square, horizontal and vertical, and old and young. Nonmathematical activities include finding and naming objects of different colors or whose first letter is a different letter from A to Z. The final scenes include a maze and a search—"Where is Mary?" Throughout the book, the reader will be involved in discriminating and classifying; the detail should intrigue the reader through many readings.

Stienecker, David L. *Addition.* Illustrated by Richard Maccabe. New York: Benchmark Books, 1996.

See Series and Other Resources: Discovering Math

Talbot, John. *It's for You: An Amazing Picture-Puzzle Book.* New York: Dutton Children's Books, 1995.

ISBN 0-525-45402-0 $16.99

★★★ Single concept, 5+

Unwrap the adventure—the opening pages depict a mysterious package wrapped in brown paper and labeled "It's for You." As the parcel is unwrapped, the reader is warned not to become "lost in the A-MAZ-ING views / Just solve the riddles to work out the clues." Riddles and puzzles, similar to those in *Games* magazine, are introduced through a text written in rhyming form; the attractive illustrations are as mysterious as the prose. By carefully analyzing both the picture and the prose, the reader can solve each challenge. One puzzle requires reading the clues in a mirror, but the reader still needs to use the clues to solve the puzzle. Another challenge involves spatial sense; it is like working a three-dimensional jigsaw puzzle. The appropriate pieces are there, but the reader has to determine which jagged pieces fit where and how they should be rotated and positioned to form a castle's tower. Other picture-puzzles involve labyrinths. A simpler puzzle involving dice and a coin is a board game disguised in a pinball-game format. Clues are given at the book's end, but even these need to be carefully considered. Most of the puzzles are not obvious; some are easier than others. At times it is difficult to determine what the question *is*. But all are brilliantly written and beautifully designed. This absorbing book will fascinate readers for hours while they analyze different aspects. Also interesting is the information about the author—an intriguing individual.

Time-Life for Children Staff. *The House That Math Built: House Math.* Alexandria, Va.: Time-Life for Children, 1993.
See Series and Other Resources: I Love Math

————. *How Do Octopi Eat Pizza Pie: Pizza Math.* Alexandria, Va.: Time-Life for Children, 1993.
See Series and Other Resources: I Love Math

————. *Play Ball: Sports Math.* Alexandria, Va.: Time-Life for Children, 1993.
See Series and Other Resources: I Love Math

————. *Pterodactyl Tunnel: Amusement Park Math.* Alexandria, Va.: Time-Life for Children, 1993.
See Series and Other Resources: I Love Math

————. *Right in Your Own Backyard: Nature Math.* Alexandria, Va.: Time-Life for Children, 1993.
See Series and Other Resources: I Love Math

Vorderman, Carol. *How Math Works.* Pleasantville, N.Y.: Reader's Digest Association, 1996.
See Series and Other Resources: Resources

White, Laurence B., and **Ray Broekel.** *Math-a-Magic: Number Tricks for Magicians.* Illustrated by Meyer Seltzer. Morton Grove, Ill.: Albert Whitman & Co., 1990.
ISBN 0-8075-4994-0 $12.95
★★★ Multiconcept, 3–6

Want to bet a million dollars that you can't fold a newspaper in half ten times? This query leads to a good discussion on the power of doubling. Twenty magic tricks using numbers are described in this 48-page reader-friendly book of mathematical recreations. After a description of each trick, instructions on how to do the trick are followed by an explanation of why the trick works—"The Math-a-Magic Secret." Mind-reading tricks that involve predicting a secret number are simply and carefully explained algebraically through diagrams. "School Daze" documents why a student has no time for homework by subtracting from 365 days such factors as hours spent sleeping, weekends, summer vacation, and school holidays. Many of the tricks involve counting and evens and odds; classic problems such as NIM, palindromes, and twelve people in eleven beds are also included. Tricks are presented with a flair that should appeal to the readers. "The Black Hole" requires a rearrangement of four puzzle pieces so that the hole in the rectangle disappears when it is reassembled. Bold print on one side of the rectangle states "THE BLACK HOLE"; the flip side of the paper states "WOW! YOU MADE THE BLACK HOLE VANISH!" The second rectangle is smaller because the arrangement does not contain a hole. Appealing drawings that illustrate the problem or reflect children solving the problems are done in red and black. In the foreword, the authors suggest that patter—entertaining talk—is part of a magician's act; it diverts the audience from the essence of the trick. This humor is often included in the illustrations.

Measurement

*B*ut the bed was too short for the queen.... The king was displeased and sent the carpenter's apprentice to jail.... The little apprentice pondered the dilemma from his prison cell.... He had carefully measured the requested dimensions in constructing the queen's bed.... What could have gone wrong?

The first graders listened intently as their teacher, Cheryl Lubinski, read Myllar's *How Big Is a Foot?* The children giggled at the picture of the tall, crowned queen lying in the little bed. They whispered to each other as they realized that the misunderstanding was due to the different sizes of the king's foot and the apprentice's foot. The children were pleased that they had solved the dilemma before the adults in the book did. The teacher then reread the book to the class. This time she asked them questions about the story, and they shared their ideas. After the discussion, they traced around their own feet on a sheet of paper and used their paper footprints to measure different lengths in the class. After the children measured the length of the room, the teacher showed them a cutout of *her* footprint. Would they get the same answer for the room's width if they used her footprint instead of one of theirs? Why? How many teacher footprints long would they estimate the room to be?

Myllar's *How Big Is a Foot?* is one of many quality children's books on measurement. Many books in the measurement category deal with time and money. These two topics can be related to subject areas other than mathematics, so they appear in two distinct sections. Because of the number and diversity of the books on time, they are further divided into three sections: telling time, concepts of time, and extensions and connections. It was not necessary to subdivide the section on money. The remaining measurement books fit into two more sections. "Beginning Measurement" deals with attributes and perceptions. "Measurement Concepts and Systems" centers on concepts and applications that are appropriate for the individual child as well as for groups in the classroom.

Beginning Measurement

Becoming aware of the different attributes of an object is essential to being able to determine what aspect to measure. Comparing relative sizes is the basis of a strategy that uses comparison to estimate the measure of an object. Exploring the attributes of an object and making comparisons are children's first introductions to measurement. Consequently, books that encourage this type of activity are classified together. Among the simplest books in this category are those that focus on contrasting attributes, such as Allington's *Opposites* and Hoban's *Push, Pull, Empty, Full*. Comparing the different attributes through examples and illustrations is the primary focus of some authors. Hoban, MacKinnon, and Pluckrose use photographs of the world around us to depict comparisons of several attributes, such as length, capacity, and weight. Miller and Wouters combine fantasy and humor to illustrate changes in size. Additional titles that also include more sophisticated ideas about comparisons are in the next category.

Allington, Richard L. *Opposites.* Illustrated by Eulala Conner. Milwaukee, Wis.: Raintree Publications, 1979.

ISBN 0-8172-1279-5   (library binding) $9.95
ISBN 0-8114-8237-5   (paper)   3.95

★★ Single concept, K–2

Fourteen pairs of opposites appear in text and in appealing pictures of familiar objects. Measurement opposites, such as big or little and tall or short, are included as well as such other opposites as many or few, same or different, and left or right. Each pair is illustrated in two similar pictures and two descriptive sentences to show the contrasts and the similarities. The book ends with an activity to review all the pairs of opposites by matching each word with its opposite and with a suggestion for making a personal book of opposites.

163

_____. *Shapes.* Illustrated by Lois Elhert. Milwaukee, Wis.: Raintree Publications, 1979.
See Geometry and Spatial Sense: Shapes

Carle, Eric. *The Grouchy Ladybug.* New York: HarperCollins Publishers, 1992.
See Measurement: Measurement of Time—Telling Time

Disney's Pop-Up Book of Opposites. New York: Disney Press, 1991.
ISBN 1-56282-018-4 $6.95
★★ Single concept, no age given

A vacation in the woods starring Goofy, Pluto, Mickey, Minnie, and Mickey's two nephews sets the scene for this book of opposites. The concepts of opposites are cleverly incorporated into the pop-ups and colorful, appealing illustrations in this sturdy, slender book. The reader can open and close the flap on Goofy's tent. By moving a tab, the reader can put Mickey either inside or outside the cabin. As a page is opened, Pluto runs under a log as Mickey jumps over it. Some other opposites include long and short, out and in, up and down, and big and little.

Felix, Monique. *The Opposites.* Mannkato, Minn.: Creative Editions, 1993.
See Geometry and Spatial Sense: Beginning Spatial Concepts

Greenfield, Eloise. *Big Friend, Little Friend.* Illustrated by Jan Spivey Gilchrist. New York: Black Butterfly Children's Books, 1991.
ISBN 0-86316-204-5 $5.95
☆ Single concept, PS–1

Everyday scenes and situations illustrate the relationships between an African American boy and his two friends—his big friend and his little friend. The contrast between big and little appears on each pair of pages. One page depicts the child being taken to a playground by his big friend, and the opposite page shows the boy and his little friend playing at home. Two illustrations effectively show the size relationship among all three friends. This beginning concept book has sturdy cardboard pages.

Hoban, Tana. *Exactly the Opposite.* New York: Greenwillow Books, 1990.
ISBN 0-688-08861-9 $15.00
ISBN 0-688-08862-7   (library binding)   14.93
★★ Single concept, PS–2

More than one comparison is implied in each pair of Hoban's stunning photographs. A photograph of a youth lying in a hammock on an autumn day is contrasted with a second photograph that shows an empty hammock at a distance. Full and empty or near and far could be represented, but the reader is to find the opposites. Other comparisons that readers can find in the natural scenes include front and back, left and right, big and little, full and empty, open and closed, and hot and cold.

_____. *Is It Larger? Is It Smaller?* New York: Greenwillow Books, 1985.
ISBN 0-688-04027-6 $14.95
ISBN 0-688-04028-4   (library binding)   15.93
★★★ Single concept, PS–1

Another excellent collection of Tana Hoban's gorgeous color photographs clearly illustrates the concept being considered. In any illustration, numerous comparisons can be made. One photograph shows two measuring cups, two bowls, three spoons, two whisks, and two spatulas. Another page shows three vases filled with three different kinds of flowers; the reader must first decide which objects are to be compared, such as the vases, before ordering them from tallest to shortest or shortest to tallest. Other attributes of the vases, such as volume, could be considered. In this no-text book, the photographs invite readers to make as many comparisons as they can. Because of the variety of everyday situations depicted, the readers should be motivated to start making comparisons in the world around them.

_____. *Push, Pull, Empty, Full: A Book of Opposites.* New York: Macmillan Publishing Co., 1972.
ISBN 0-02-744810-X $13.95
★★★ Single concept, PS–2

Hoban makes another successful photographic journey in the exploration of the concept of opposites. Her black-and-white photographs of the strikingly familiar environment that surrounds the city child provide fifteen pairs of opposites (push, pull; empty, full; in, out; thick, thin; whole, broken; etc.). Even the end pages of this well-designed book reflect the theme of opposites because the beginning shows a fully leafed tree and the end depicts the naked branches of the same tree against the wintry sky.

Inkpen, Mick. *Kipper's Book of Opposites.* San Diego, Calif.: Red Wagon Books, Harcourt Brace & Co., 1994.
ISBN 0-15-200668-0 $6.00
☆ Single concept, PS

Kipper, a brown and white puppy, introduces eight pairs of opposites in this small book. Kipper chases a gray squirrel that climbs "up" a tree and then scurries "down" the tree on the opposite page. For another pair of opposites, Kipper is shown hugging a big teddy bear while observing a very small teddy bear on the opposite page. The appealing figures are simply illustrated against a white background. The minimal text includes only the names for each opposite pair. Other opposites include new, old; out, in; happy, sad; slow, fast; long, short; and day, night.

MacKinnon, Debbie. *What Size?* Illustrated by Anthea Sieveking. New York: Dial Books for Young Readers, 1995.

ISBN 0-8037-1745-8 $10.99

★★ Single concept, PS–K

Similar in format to MacKinnon's *How Many?* and *What Shape?*, this book incorporates beautiful photographs of charming children in familiar settings to illustrate comparative size in a meaningful way. Each double-page spread is designed the same way, with the left-hand page focusing on children in a natural setting. One example includes "Ezra and Kelly are playing dress-up. Kelly's dress is the longest." The colorfully bordered opposite page, headed "Long and Short," shows three dresses of various lengths and poses the question, "Which dress is the shortest?" Big, little; thick, thin; high, low; wide, narrow; and tall, small are all opposites that are similarly dealt with. A discussion of same size and different size leads into an invitation for the children to find out how tall they are. A crisp format is achieved by a simple text printed in large type and close-up photographs with a white background. Additionally, children from different ethnic backgrounds have been included throughout this series.

McMillan, Bruce. *Becca Backward, Becca Frontward.* New York: Lothrop, Lee & Shepard Books, 1986.

See Geometry and Spatial Sense: Beginning Spatial Concepts

Miller, Ned. *Emmett's Snowball.* Illustrated by Susan Guevara. New York: Henry Holt & Co., 1990.

ISBN 0-8050-1394-6 $14.95

★★ Single concept, PS–2

Emmett rises early one winter morning to play in the fresh snow. He builds a snowball and rolls it around his yard until it is half as tall as he is. Just as he is running out of snow, his friend Sarah joins him and they push the snowball through Sarah's yard to build a snowball as tall as Sarah. The story line continues with more of the townspeople participating as the snowball continues to increase in size. The process is reversed when the weather becomes warm and Emmett's snowball is reduced to its original size. Size comparisons are made throughout as the big snowball is considered. The humorous story will appeal to the young reader. Pleasing pastel illustrations capture the children's expressions and antics. Additionally, the cast of friends includes children from various racial backgrounds.

Moncure, Jane Belk. *The Biggest Snowball of All.* Illustrated by Joy Friedman. Chicago: Children's Press, 1988.

ISBN 0-89565-391-5 $21.36
ISBN 0-89565-918-2   (paper)   21.36

☆ Single concept, PS–2

Little Bear's tiny snowball rolls on and on, gathering more snow and becoming the biggest snowball of all. The story clearly illustrates comparative terms from *tiny* to *small, big, bigger, even bigger,* and *biggest of all.* Some of the terms are reinforced when three friends hop on a sled with "the smallest in front and the biggest in back." The idea of a hollow snowball, when "Little Bear made a wide door so they could get inside," is most unbelievable and will probably confuse the young reader. When the sun comes out, the biggest snowball of all melts down to the tiny snowball it once was. Simple pastel illustrations help tell the story.

Murphy, Stuart J. *The Best Bug Parade.* Illustrated by Holly Keller. New York: HarperCollins Publishers, 1996.
See Series and Other Resources: MathStart

My First Look at Sizes. New York: Random House, 1990.
ISBN 0-679-80532-X $9.00
★★ Single concept, PS–K

Comparisons and sequences are effectively illustrated in the crisp, colorful photographs. Eight various-sized balls are sequenced from little to big, whereas nine seashells are arranged from big to little. Three decorative gifts depict big, bigger, and biggest, and Russian dolls illustrate small, smaller, and smallest. Carefully selected round things—an apple, a tomato, an orange, a ball of yarn, and an onion—clearly depict objects of the same size. Same size is also shown by considering each object in a pair, such as boots, gloves, and socks. Appealing illustrations and simple text dominate this thoughtfully designed beginning book.

Pienkowski, Jan. *Sizes.* New York: Little Simon, 1990.
ISBN 0-671-72844-X $2.95
★★★ Single concept, PS–K

Size is clearly depicted as comparison in this extremely colorful book; the concepts of big and little are represented by such contrasting illustrations as an ocean liner and a little sailboat and a tree and a flower. Minimal text accompanies the brightly colored graphics. Originally published in 1973, this Pienkowski book is reprinted as a nursery board book in an 8.5 cm by 8.5 cm format.

Pluckrose, Henry. *Size.* Illustrated by Chris Fairclough. Chicago: Children's Press, 1995.
See Series and Other Resources: Math Counts

Reiser, Lynn. *Christmas Counting.* New York: Greenwillow Books, 1992.
See Early Number Concepts: Counting—One to Ten

Rogers, Paul. *Surprise, Surprise!* Illustrated by Sian Tucker. New York: Scholastic, 1990.
See Geometry and Spatial Sense: Beginning Spatial Concepts

Spier, Peter. *Fast-Slow, High-Low: A Book of Opposites.* New York: Doubleday & Co., 1988.

ISBN 0-385-24093-7 $5.95

☆ Single concept, K–3

The opposites big and small are illustrated through such pairs as circus tent and pup tent, ostrich and robin, city and town, and whale and fish. Each concept is named and then illustrated with a colorful collage of eight to fifteen examples. Young and old, old and new, long and short, deep and shallow, up and down, straight and crooked, on and off, tall and short, open and closed, full and empty, heavy and light, wide and narrow, same and different, more and less, and few and many are some of the concepts that are effectively illustrated by Spier. The closing task is front-back, top-bottom; the examples range from a colored pencil and a child on a tricycle to a farmer standing in line with a dog, a pig, a chicken, a horse, and a cow. The reader should enjoy these humorous examples while exploring spatial perception.

Tucker, Sian. *Sizes.* New York: Little Simon, 1992.

ISBN 0-671-76909-X $2.95

☆ Single concept, PS–K

Bold color dominates the backgrounds and the objects to be counted in this tiny cardboard-paged book. Size comparisons, such as little shoes, big shoes; little building, big building; and little car, big truck, are usually shown on opposite pages. In a different type of example, the tiny ladybug–small mouse comparison is followed by a two-page spread of an enormous dinosaur. Similarly,the little butterflies–big bird comparison is followed by a huge airplane. The last page shows clothing hung on a line with the caption "all different sizes."

Wouters, Anne. *This Book Is Too Small.* New York: Dutton Children's Books, 1992.

ISBN 0-525-44881-0 $8.95

★★ Single concept, PS–1

Each page that depicts Bear has a green background lightly framed in black. Opening pages show baby Bear, who grows bigger and bigger on each consecutive page! His discomfort becomes increasingly apparent because "this book is too small" and its page cannot adequately contain Bear. But Little Mole comes to the rescue by erasing part of the black line that encloses Bear on the page. With Little Mole's help, Bear is able to squeeze to freedom. The two friends run off together, their shapes diminishing in the distance. Without text and with scarcity of line, the simple, expressive illustrations clearly convey a story that focuses on relative size.

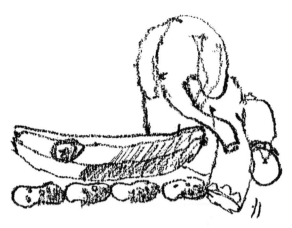

Measurement Concepts and Systems

The attributes of length, weight, area, volume, temperature, and angles are explored in greater depth in this second measurement category. Arbitrary units of measurement, nonstandard and standard, are used to experiment with the process of measuring and to introduce measurement systems. Adam's unique book *Ten Beads Tall* will give young readers many opportunities to work with an arbitrary unit of measurement, and Myllar's *How Big Is a Foot?* effectively relays the need for a standard unit. Comparing, measuring, and collecting data about one's self is the focus in *From Head to Toe: Body Math;* additional measurement activities are included in the other books authored by the Time-Life for Children Staff. Measurement of liquids and solids is introduced in Pluckrose's *Capacity;* in the Math Counts series, Pluckrose's real-life examples move from attributes to comparisons to measurements. Fascinating comparisons found in the work by Ash, Norden and Ruschak, Schwartz, and Wells integrate measurement, number, and data. Finally, more sophisticated real-life applications of measurement can be found in Markle's *Measuring Up!: Experiments, Puzzles, and Games Exploring Measurement* as well as in the detailed works by Smoothey and Walpole. Smoothey's works include a number of investigations for the older reader as well as extensions to formula development.

The following book are out of print but are considered exceptional:

Branley, Franklin M. *How Little and How Much: A Book about Scales*

———. *Measure with Metric*

Fey, James. *Long, Short, High, Low, Thin, Wide*

Froman, Robert. *Bigger and Smaller*

Laithwaite, Eric. *Size: The Measure of Things*

Linn, Charles F. *Estimation*

Srivastava, Jane Jonas. *Area*

———. *Spaces, Spaces, and Sizes*

———. *Weighing and Balancing*

Copies may be available in your library. See the first edition of *The Wonderful World of Mathematics* for annotations.

Adams, Pam. *Ten Beads Tall.* Sudbury, Mass.: Child's Play International, 1989.
ISBN 0-85953-242-9 $11.99

★★ Single concept, PS–2

Sturdy cardboard pages, bright primary colors, and a string of multicolored beads nestling in a long furrow gouged in the top of the book may fool the reader into believing that this is a toy book for a preschooler. Not so! This unique little book is compelling for the youngster who is beginning to explore the concepts of size relationships and measurement. The string of ten multicolored cubes is the tool that helps readers accept the challenge to measure the lengths and widths of objects on each page. "How tall?" "How wide?" "How long?" "How tall and high?" "How long, high, and wide?" are some of the questions asked. The last two pages offer special challenges. On one page, a variety of vehicles approach a tunnel, and the reader is asked to determine which ones will pass through. On the final page the reader is asked to use the measure to match various toys with the appropriate storage box.

Allen, Pamela. *Mr. Archimedes' Bath.* New York: Lothrop, Lee & Shepard Books, 1980.
ISBN 0-207-17285-4 $7.95

★★ Single concept, PS–2 (K–3)

Pen-and-ink lines and watercolors create the charming illustrations throughout this humorous story of Mr. Archimedes' discovery of measuring volume by water displacement. This improbable tale shows Mr. Archimedes pondering why his bath always overflows. He measures the depth of the water but finds that as he, a kangaroo, a goat, and a wombat get into the tub, the water flows over the rim. They experiment by getting in and out of the tub, each time measuring the level of the water. Mr. Archimedes orders his friends out one by one. "Eureka!" He finally discovers that *they* make the water level go up and down. This delightful tale is a good introduction to water displacement by objects.

———. *Who Sank the Boat?* Sandcastle Books, 1990.
ISBN 0-698-20679-7 $5.95

★★ Single concept, (K–3)

Originally published in 1982, this humorous book was reissued in paper by Sandcastle. Like those in Allen's *Mr. Archimedes' Bath,* the charming illustrations were created with pen-and-ink lines and watercolors. The question is, Which of five friends—a cow, a donkey, a sheep, a pig, and a mouse—sank the boat? One by one, each character is shown as he or she precariously get into the boat. The waterline approaches the rim of the boat, but not until the final scene does the reader see the boat sink as the little mouse climbs on board. This delightful tale is a good introduction to the concepts of volume and capacity.

Anno, Mitsumasa. *Anno's Math Games.* New York: Philomel Books, 1987.
See Number—Extensions and Connections: Number Theory, Patterns, and Relationships

———. *Anno's Math Games II.* New York: Philomel Books, 1989.
See Number—Extensions and Connections: Number Theory, Patterns, and Relationships

Ash, Russell. *Incredible Comparisons.* New York: Dorling Kindersley, 1996.
See Number—Extensions and Connections: Statistics and Probability.

Axelrod, Amy. *Pigs in the Pantry.* Illustrated by Sharon McGinley-Nally. New York: Simon & Schuster Books for Young Readers, 1997.
ISBN 0-689-80665-5 $13.00
☆ Single concept, PS–4

Fun with Math and Cooking is the subtitle of another humorous adventure of the Pig family. Because Mrs. Pig is not feeling well, the family insists that she stay in bed while they prepare her favorite dish, firehouse chili. After briefly consulting the recipe (the ingredients, methods, and variations are included on a two-page spread), the family begins cooking with relish. As Mr. Pig adds the ingredients to the overflowing pots, the reader sees that he has mixed up the measures of different ingredients and that he does not have a sense of the reasonableness of measures or ingredients as he adds one carrot, eight cloves of garlic, two-thirds cup of salt, and a cup and a half of pepper. The story ends with the arrival of the fire department to investigate the smoke escaping from the kitchen. At the end of the book are measurement equivalents for cooking; words that cooks use, such as *chop, mince,* and *recipe;* and questions to the reader about the mistakes Mr. Pig made while cooking. Axelrod's humorous story of the Pig family is again complemented with McGinley-Nally's colorful, lively illustrations.

Bulloch, Ivan. *Measure.* New York: Thomson Learning, 1994.
See Series and Other Resources: Action Math

Burns, Marilyn. *The I Hate Mathematics! Book.* Illustrated by Martha Hairston. Boston: Little, Brown & Co., 1975.
See Number—Extensions and Connections: Number Theory, Patterns, and Relationships

Clement, Rod. *Counting on Frank.* Milwaukee, Wis.: Gareth Stevens Children's Books, 1991.
See Number—Extensions and Connections: Number Concepts

Griffiths, Rose. *Boxes.* Illustrated by Peter Millard. Milwaukee, Wis.: Gareth Stevens Publishing, 1995.
See Series and Other Resources: First Step Math

———. *Facts & Figures.* Illustrated by Peter Millard. Milwaukee, Wis.: Gareth Stevens Publishing, 1994.
See Series and Other Resources: First Step Math

Hightower, Susan. *Twelve Snails to One Lizard: A Tale of Mischief and Measurement.* Illustrated by Matt Novak. New York: Simon & Schuster Books for Young Readers, 1997.
ISBN 0-689-80452-0 $13.00
☆ Single concept, PS–2

Bubba Bullfrog sprawls comfortably along the water's edge as Milo Beaver raises the alarm that the pond may be dry by summer because of the break in the dam. Bubba acknowledges Milo's concern and condescends to think about, but not to work physically on, the problem of how to measure branches to patch the hole. Because Milo has been finding branches too long or too short, Bubba volunteers the information that the dam break is thirty-six inches. But Milo has no idea what an inch is. When told that a snail is an inch long, Milo rounds up thirty-six snails to determine the needed length. Subsequent illustrations show Milo patiently trying to encourage thirty-six snails to stand in line. Similar episodes continue as he tries to line up three foot-long lizards or to coax Betty Jo Boa to stretch out her yard-long length. When all these trials fail, Bubba presents a yardstick, subdivided in inches and feet, to a harried Milo. This book succeeds in introducing customary length measures and the relationships among such measures in a humorous manner. The pastel illustrations enhance the exaggerated story line in this mischievous tale. A final page includes measurement facts about the animals mentioned in the book.

Lasky, Kathryn. *The Librarian Who Measured the Earth.* Illustrated by Kevin Hawkes. Boston: Little, Brown & Co., 1994.
See Geometry and Spatial Sense: Geometric Concepts and Properties.

Markle, Sandra. *Math Mini-Mysteries.* New York: Atheneum, 1993.
See Number—Extensions and Connections: Number Theory, Patterns, and Relationships.

————. *Measuring Up!: Experiments, Puzzles, and Games Exploring Measurement.* New York: Atheneum Books for Young Readers, 1995.

ISBN 0-689-31904-5 $17.00
ISBN 0-689-80618-3   (Aladdin Paperbacks)   7.95

★★ Single concept, 4–7

Magnifications of a strawberry clearwing moth and a pink cricket, optical illusions, unique measuring tools like a clinometer and a hodometer, and the height of the tallest person who played in the National Basketball Association are featured in activities for measuring lengths. In addition to a chapter on length and size, other sections include information and activities on quantities; weights; temperatures; and volume, perimeter, and area. Measuring quantities includes the collecting data as well as baking fortune cookies. Designing and conducting experiments for keeping tea warm are part of measuring temperatures. An interesting puzzle about weights has multiple solutions and is based on the story of Jack and the beanstalk. By the same author as *Math Mini Mysteries, Measuring Up* has many of the same strengths: hands-on activities; problem solving and explorations; experiments, puzzles, and games. However, this book focuses on one concept—measurement—with the context of situations relevant to children. Additionally, its eye-catching color photography and less dense print make this book much more attractive and appealing. Three elementary children from different ethnic backgrounds are shown actively engaged in the many experiments. Solutions, an index, and a table of contents are included. An opening section states that the tasks are designed to support NCTM's *Curriculum and Evaluation Standards for School Mathematics.*

Massin. *Fun with 9umbers.* Illustrated by Les Chats Pelés. San Diego, Calif.: Creative Editions, 1993.

See Number—Extensions and Connections: Number Systems

My First Number Book. New York: Dorling Kindersley, 1992.

See Number—Extensions and Connections: Number Concepts

McKibbon, Hugh William. *The Token Gift.* Illustrated by Scott Cameron. New York: Amick Press, 19969.

See Number—Extensions and Connections: Number Theory, Patterns, and relationships

Myllar, Rolf. *How Big Is a Foot?* New York: Dell Publishing, 1991.

ISBN 0-440-40495-9   (paper) $3.50

★★★ Single concept, PS–3

As a surprise for the queen's birthday, the king decides to have a bed made for her. (Beds haven't been invented yet; so it will really be a surprise.) The large king marks off the dimensions for the proposed bed with his feet. He gives these dimensions to the head carpenter who gives them to his little apprentice. When the bed is delivered, it is too small, and the apprentice is jailed. From his jail cell, the apprentice solves the problem, and everyone lives happily ever after. This humorous tale about nonstandard measures will delight young readers. Simple drawings illustrate this narrative.

Norden, Beth B. and **Lynette Ruschak.** *Magnification: A Pop-up Lift-the-Flap Book.* Illustrated by Wendy Smith-Griswold, Matthew Bens, and Robert Hynes. New York: Lodestar Books, 1993.

ISBN 0-525-67417-9 $14.99

★★ Single concept, 2+

Such everyday items as table salt, record grooves and dust bunnies become unrecognizable when they are magnified. A gram of table salt appears as a large two-inch cube after it has been magnified 144 times. A piece of pink string, attached to the page, is shown magnified 20 times; as a flap is turned, two views showing magnifications of 150 times and then 1000 times appear! The results are dramatic and could be used to introduce the topics of similarity and scalar relationships. Sturdy pop-ups and lift-the-flaps contribute to the surprises throughout the book. Connections are made to why and how we use magnifications. For example one page relates how medical researchers examine tissues to understand better how cells function and change with age or disease; magnifications include viruses and dental plaque. A human hair magnified 1516 times looks like a plant stem. At 130 times larger than life, a cardboard pop-up flea dominates the page exemplifying the complexity and diversity of insects.

Pluckrose, Henry. *Capacity.* Illustrated by Chris Fairclough. Chicago: Children's Press, 1995.
See Series and Other Resources: Math Counts

———. *Length.* Illustrated by Chris Fairclough. Chicago: Children's Press, 1995.
See Series and Other Resources: Math Counts

———. *Numbers.* Illustrated by Chris Fairclough. Chicago: Children's Press, 1995.
See Series and Other Resources: Math Counts

———. *Weight.* Illustrated by Chris Fairclough. Chicago: Children's Press, 1995.
See Series and Other Resources: Math Counts

Ross, Catherine Sheldrick. *Circles: Fun Ideas for Getting A-Round in Math.* Illustrated by Bill Slavin. Reading, Mass.: Addison Wesley, 1992.
See Geometry and Spatial Sense: Geometric Concepts and Properties.

Schwartz, David M. *How Much Is a Million?* Illustrated by Steven Kellogg. New York: Lothrop, Lee & Shepherd Books, 1985.
See Number—Extensions and Connections: Number Concepts

————. *If You Made a Million.* Illustrated by Steven Kellogg. New York: Lothrop, Lee & Shepherd Books, 1989.
See Measurement: Money

Scieszka, Jon. *Math Curse.* Illustrated by Lane Smith. New York: Viking, 1995.
See Number—Extensions and Connections: Number Concepts

Smoothey, Marion. *Area and Volume.* Illustrated by Ted Evans. New York: Marshall Cavendish, 1992.
See Series and Other Resources: Let's Investigate

————. *Estimating.* Illustrated by Ann Baum. New York: Marshall Cavendish, 1995.
See Series and Other Resources: Let's Investigate

————. *Time, Distance, and Speed.* Illustrated by Ted Evans. New York: Marshall Cavendish, 1993.
See Series and Other Resources: Let's Investigate

Time-Life for Children Staff. *The Case of the Missing Zebra Stripes: Zoo Math.* Alexandria, Va.: Time-Life for Children, 1992.
See Series and Other Resources: I Love Math

————. *From Head to Toe: Body Math.* Alexandria, Va.: Time-Life for Children, 1993.
See Series and Other Resources: I Love Math

————. *The House That Math Built: House Math.* Alexandria, Va.: Time-Life for Children, 1993.
See Series and Other Resources: I Love Math

————. *How Do Octopi Eat Pizza Pie: Pizza Math.* Alexandria, Va.: Time-Life for Children, 1993.
See Series and Other Resources: I Love Math

————. *The Mystery of the Sunken Treasure: Sea Math.* Alexandria, Va.: Time-Life for Children, 1993.
See Series and Other Resources: I Love Math

————. *Play Ball: Sports Math.* Alexandria, Va.: Time-Life for Children, 1993.
See Series and Other Resources: I Love Math

————. *Right in Your Own Backyard: Nature Math.* Alexandria, Va.: Time-Life for Children, 1993.
See Series and Other Resources: I Love Math

————. *See You Later Escalator: Mall Math.* Alexandria, Va.: Time-Life for Children, 1993.
See Series and Other Resources—I Love Math

Vorderman, Carol. *How Math Works.* Pleasantville, N.Y.: Reader's Digest Association, 1996.
See Series and Other Resources: Resources

Walpole, Brenda. *Distance.* Illustrated by Dennis Tinkler and Chris Fairclough. Milwaukee, Wis.: Gareth Stevens, 1995.
See Series and Other Resources: Measure Up with Science

————. *Size.* Illustrated by Dennis Tinkler and Chris Fairclough. Milwaukee, Wis.: Gareth Stevens, 1995.
See Series and Other Resources: Measure Up with Science

————. *Speed.* Illustrated by Dennis Tinkler and Chris Fairclough. Milwaukee, Wis.: Gareth Stevens, 1995.
See Series and Other Resources: Measure Up with Science

————. *Temperature.* Illustrated by Dennis Tinkler and Chris Fairclough. Milwaukee, Wis.: Gareth Stevens, 1995.
See Series and Other Resources: Measure Up with Science

Wells, Robert E. *Is a Blue Whale the Biggest Thing There Is?* Morton Grove, Ill.: Albert Whitman & Co., 1993.

ISBN 0-8075-3655-5 (library binding)	$13.95
ISBN 0-8075-3656-3 (paper)	6.95

★★ Single concept, 1–6

Developing number sense and an appreciation of the comparative size of parts of our physical world is the focus of this slender book. A first comparison involves the fluke, the flipper part of a blue whale's tail that is bigger than most of Earth's creatures. How much bigger is shown by including in the illustration the comparative sizes of a horse, an elephant, and a lion who are looking on from the pier. This trio looks even smaller as the entire blue whale is shown. The whale's statistics, such as a length of 100 feet and a weight of 150 tons, are most impressive. The query "Is a blue whale the biggest thing there is?" generates new comparisons. Next to two really big jars each containing 100 blue whales, the trio looks extremely small; but as the jars are stacked to form a tower 10 jars high, the trio becomes tiny. But the huge tower looks like a banner when it is shown on top of Mount Everest, and the trio can no longer be seen. One hundred Mount Everests stacked on top of each other are graphically depicted as "a mere WHISKER on the face of the Earth," but 100 Earths packed into a big bag are small when pictured next to the sun. The simple, colorful illustrations effectively show the comparisons being made; the examples are excellent and these ratios help children begin to contemplate the size of our universe.

————. *What's Smaller than a Pygmy Shrew?* Morton Grove, Ill.: Albert Whitman & Co., 1995.

ISBN 0-8075-8837-7 $13.95
ISBN 0-8075-8838-5 (paper) 6.95
☆ Single concept, 1–6 (3–6)

Like Wells's *Is a Blue Whale the Biggest Thing There Is?* this book focuses on an appreciation of the comparative size of different things in our world. A first comparison involves a pygmy shrew who at three inches "from the end of her nose to the tip of her tail" looks small even standing next to a toadstool. And she may be the smallest thing in the universe when compared with an elephant. This leads to a comparison to show that the pygmy shrew is not so small when compared with a ladybug. The ladybug is not so small when compared with protozoa living in the water on a leaf. Each new comparison is reflected in the illustrations that show the relative sizes; more detail is noted as the objects are enlarged for comparisons with smaller objects. Next to a pygmy shrew the spotted wings of a ladybug are clearly seen, but the "creepy" details of the ladybug's eyes, feelers, and legs appear when it is depicted next to protozoa. The magnifications continue with colored illustrations of amoebas and paramecia and extend to molecules, atoms, protons, neutrons, and quarks. The text includes information describing each new entity because the major focus of this book is on physics and nuclear size; relative size is a secondary focus. The simple, colorful illustrations effectively show the comparisons being made, and the relative sizes help a reader begin to contemplate the minute components of matter.

Ziefert, Harriet. *Measure Me: A Counting Book and Growth Chart.* Illustrated by Susan Baum. New York: Harper Collins, 1991.

ISBN 0-694-00322-0 $12.95
★★ Single concept, PS–3

One by one, five animals line up across the successive pages; each new animal is bigger than the previous ones. Large black numerals record the number in each increasing set. The text parallels the comparisons shown in the illustrations: "A cow is bigger than a mouse, and a duck, and a pig." After a giraffe is introduced, the reader is asked, "How big are you?" and the height chart is introduced. The card-stock pages unfold to a chart five feet long. Each page is one foot long and includes the appropriate numeral and word name. A final page is designed for recording the reader's height and for attaching photographs. This well-designed book cleverly links counting objects and comparing sizes to measuring lengths.

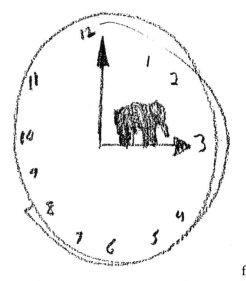

Measurement of Time— Telling Time

Learning to tell time both on face and digital clocks is the introduction to time concepts for most children. The simplest books in this category relate time to the hour, such as Dijs's *What Do I Do at 8 O'Clock?* and Pienkowski's *Time*. Events often reflect the child's day and include invitations for involvement. A clock face with movable hands is embedded in books by Anastasio, Katz, and Llewellyn; a watch is attached to Cassidy's *The Time Book*. These books as well as those by Viorst and others introduce concepts of time to the quarter and half hour, five-minute intervals, and before and after the hour. Other dimensions of telling time are seen in the puzzlelike illustrations in Edens's *The Wonderful Counting Clock* and the poetry selections by Lee Bennett Hopkins in *It's About Time!*

Anastasio, Dina. *It's About Time.* Illustrated by Mavis Smith. New York: Grosset & Dunlap, 1993.

ISBN 0-448-40551-2 $9.95

☆ Single concept, PS+ (2–4)

Movable black plastic hands on a large watch face appear through the circular cutout on the book's cover. The clock's face always appears through a right-hand page as the story progresses. Tasks are posed throughout the book to engage the reader in setting and telling time to the hour, to the half hour, and to five-minute intervals. The story line traces the day's activities of Tim and his dog Ticker. Directions are given for moving each hand to correspond to the time for a particular activity. To elucidate the relationship of sixty minutes in an hour, the reader is asked to move the minute hand one revolution and the hour hand from 8 to 9 to show the hand's passage from 8 o'clock to 9 o'clock. For telling time "after" or "to" the hour, a smaller concentric circle on the clock's face is divided into two different-colored semicircles. Near the smaller circle's circumference in smaller print are such phrases as "5 after" next to the numeral 1,

178

"20 after" next to the 4, and "20 to" next to the 8. When the minute hand is correctly placed, these words appear through an opening on the minute hand. The reader is also asked to place the hour hand halfway between two numerals to show time to the half hour and close to one of two appropriate numerals to show time "after" and "to" the hour. The clever design and placement of text make this a unique book for introducing the concept of time.

Carle, Eric. *The Grouchy Ladybug.* New York: HarperCollins Publishers, 1992.

ISBN 0-06-027087-X		$15.95
ISBN 0-06-027088-8	(library binding)	15.89
ISBN 0-06-443450-8	(paper)	6.95

★★★ Single concept, PS–1 (PS–3)

An opening page in beautiful blues and greens shows fireflies dancing in the moonlit night. Carle's collages tell the story of the grouchy ladybug's encounters during the daylight hours. Morning dawns at five o'clock and two ladybugs land on the same leaf to eat aphids. The grouchy ladybug leaves in a huff, saying that it will find someone bigger to fight. It goes on to encounter a yellow jacket at six o'clock, a stag beetle at seven o'clock, a praying mantis at eight o'clock, ..., an elephant at four o'clock; each time, the ladybug challenges the creature with, "Hey, you want to fight?" and leaves looking for someone bigger to fight. A small clock face is located in each page's upper right-hand corner; these pages are layered so that all the clock faces from six o'clock through five o'clock can be seen at once. All the times are to the hour except the last sequence, when a whale is encountered. At five fifteen the ladybug approaches the whale's flipper; at five thirty, a fin; and at quarter to six, the whale's tail. The whale's tail makes a splash that sends the ladybug back to the beginning where it reconciles with the friendly ladybug at six o'clock. Fireflies reappear, which completes the cycle of the day. The book's primary themes are telling time, the cycle of a day, behavior, and size comparisons.

Cassidy, John. *The Time Book.* Illustrated by Roger Bollen. Palo Alto, Calif.: Klutz Press, 1991.

ISBN 1-878257-08-0 $10.95

★★★ Single concept, no age given

Attached to the front of this slender, cardboard, spiral-bound book is a colorful quartz watch with a second hand. The reader can use this watch to answer the questions in the book. Good explanations with illustrations help the reader estimate time by using only the hour hand for times "on the hour," "a little after," "half past," and so on. After this introduction, telling time with the minute hand is explored. The watch's hands do not move continuously, so the reader can see the hand jump from one minute mark to the next. After learning to count the intervals by ones, the reader is introduced to a shortcut of counting by fives. Interwoven among the telling-time lessons are references to, and interesting facts about, time: "Stuff that takes about one hour" could be watching two TV programs or eating twelve ice-cream cones. The humorous illustrations and examples should appeal to readers.

Dijs, Carla. *What Do I Do at 8 O'clock?* Cambridge, Mass.: Simon & Schuster Books for Young Readers, 1993.

ISBN 1-671-79526-0 $8.95

☆ Single concept, PS–K

What do *you* do from 8 o'clock in the morning until 7 o'clock at night? A unique quality of this pop-up book on time is the cutout that allows the clear-faced clock with bright red movable hands to be seen from the front cover, through each page, until it is found on the last page. On opening this attractive pop-up book, the young reader will be greeted by the rooster who declares that he wakes everybody at 8 o'clock; the next page turn reveals the cat who says that she eats breakfast at 9 o'clock. Throughout the day, different animals state their actions at a particular time when queried about what they do at different hours. Only the cow takes more than an hour; she takes a nap from 1 o'clock to 3 o'clock. Children will enjoy the pop-up action of each of the book's characters and the opportunity to manipulate the plastic hands of the clock to match the text. Children who are beginning their experience with telling time and reading text may be confused by the "O'Clock," since the O can be mistaken for a numeral.

Dodds, Dayle Ann. *Someone Is Hiding.* Illustrated by Jerry Smath. New York: Little Simon, 1994.

See Early Number Concepts: Counting—Partitions

Edens, Cooper. *The Wonderful Counting Clock.* Illustrated by Kathleen Kimball. New York: Simon & Schuster Books for Young Readers, 1995.

ISBN 0-671-88334-8 $15.00

★★ Single concept, 1–3 (2–4)

The text, in large bold print, shares what happens at specific times between one o'clock and midnight on a unique clock tower. The alliterative and descriptive language evokes colorful images that are reinforced through Eden's magnificent, full-color illustrations. From ten past one, when the "handsome harlequin presents 1 dazzling sun," to twelve o'clock, when "proud Pierrot pledges 12 faithful hearts," the book offers opportunities for counting from one through twelve. A missing numeral on each clock face corresponds to the number to be counted. Most of the clock faces are only partially shown, so it is up to the reader to verify the time. The minute hand is not visible for three o'clock, but the hour hand points straight to the 3's location. For twenty-one to four, both the hour and minute hands are illustrated with the appropriate angles; because only part of the clock's face is visible, the reader must determine between what numbers the hands are located. The final illustration indicates one minute after twelve with all twelve guides standing on the clock tower. The text reminds the reader, "isn't it time you were asleep and dreaming?"

Füchshuber, Annegert. *The Cuckoo-Clock Cuckoo.* Minneapolis, Minn.: Carolrhoda Books, 1988.

ISBN 0-87614-320-6 $18.95

☆ Single concept, PS–3

Numerous clock faces with corresponding digital readings are shown throughout the book, depicting such times as 12:00 P.M., 4:00 A.M., 4:02 A.M., 11:37 A.M., and 4:45 P.M. Each time period traces the adventures of a cuckoo who is trying to get back into the clock after being locked out. Each illustration of the outside of the home where the cuckoo clock resides reflects the day's activities and the changing light patterns. The fanciful story line and illustrations should appeal to the young reader.

Grey, Judith. *What Time Is It?* Illustrated by Susan Hall. Mahwah, N.J.: Troll Associates, 1981.

ISBN 0-89375-509-5 $11.59
ISBN 0-89375-510-9   (paper)   2.95

☆ Single concept, K–2

"What time is it?" is the repeated refrain of the bushy squirrel to other creatures throughout the book. Each query is answered with a set of questions, such as "Is it time for breakfast? Lunch? Dinner?" or "Is it time to run? Swim? Ride?" The ultimate statement is "It is time to read!" The main purposes of the text appear to be establishing a beginning reading vocabulary and emphasizing the importance of reading. The questions could promote a discussion about various activities at different times of the day. The colorful cartoonlike illustrations are consistent with the simple text.

It's About Time! Poems selected by Lee Bennett Hopkins. Illustrated by Matt Novak. New York: Simon & Schuster Books for Young Readers, 1993.

ISBN 0-671-78512-5 $14.00

☆ Single concept, PS–3

Poems selected by the well-known anthologist Lee Bennett Hopkins have been chosen to represent appropriate activities and thoughts for specific times of the day. The connection between the charming verse and the time marked on the clock high in an upper corner of each double page is sometimes implied. Underneath each clock, the indicated time is written in numerals. Such well-known poets as Harry Behn, Aileen Fisher, Dorothy Aldis, and Gwendolyn Brooks trace the passage of one day from 7:00 A.M. to midnight.

Illustrations are pleasingly colorful and reflect the mood of individual poems.

Katz, Bobbi. *Tick-Tock, Let's Read the Clock.* New York: Random House, 1988.

ISBN 0-394-89399-9 $7.95

☆ Single concept, PS–1

A clock with movable hands and with dials to show digital times is sturdily incorporated on the last page of this book. The previous pages have cutouts so that this aid can be used for each page. The times are limited to 7:00, 8:30, 12:00, 4:00, 7:30, and 12:00. For each time, the reader is asked to change the hands of the clock and the dials so that the digital clock corresponds to the clock face. The directions are explicit; for 8:30, the short hand is to point halfway between 8 and 9 and the long hand is to point to 6. Pastel, cartoonlike illustrations on the opposite pages show children involved in appropriate activities for each time.

Llewellyn, Claire. *My First Book of Time.* New York: Dorling Kindersley, 1992.

See Measurement: Measurement of Time— Concepts

Merriam, Eve. *Train Leaves the Station.* Illustrated by Dale Gottlieb. New York: Bill Martins Books, 1994.

See Early Number Concepts: Chants, Songs, and Rhymes.

Pienkowski, Jan. *Time.* New York: Little Simon, 1990.

ISBN 0-671-72847-4 $2.95

☆ Single concept, PS–K

Simple illustrations in brilliant color combinations dominate this book that is typically Pienkowski. Each left-hand page shows a clock face. The minimal text records the time. The times are all to the hour; not all the hours between five o'clock and midnight are represented. On the opposite pages are scenes of activities appropriate for each time. From the rooster crowing at five o'clock to playtime at eleven o'clock to bathtime at six o'clock, the colorful scenes should appeal to the young reader. Originally published in Great Britain, this book has been reprinted as a nursery board book in an 8.5 cm by 8.5 cm format.

Roy, Cal. *Time Is Day.* New York: Astor-Honor, 1968.

ISBN 0-8392-3065-6 $9.95

☆ Single concept, K–3

Time elapses from 5:00 A.M. to 9:00 P.M. as a brilliant sun changes position from page to page. Colorful illustrations of a clock, the sun, and the actions of a child depict three ways to tell the time as the day progresses. Blocks of time are omitted between 9:00 A.M. and 12:00

noon and between 1:00 P.M. and 3:30 P.M. No explanation is given for this or for the sudden change from presenting only the hour to the inclusion of half hours, such as 3:30 P.M. After a cursory introduction of night and day, the text switches to years and seasons without logical development. Time concepts are not presented in a logical sequence or with adequate explanations, making the book inappropriate as an introduction to the concept of time.

Viorst, Judith. *Sunday Morning.* Illustrated by Hilary Knight. New York: Atheneum, 1992.

ISBN 0-689-31794-8 $13.95

★★ Single concept, PS–1

The frenetic energy of two little boys is captured in this delightful story of Anthony and Nicholas who have been ordered *not* to disturb their parents until 9:45 A.M. on Sunday morning. Action packed, silhouetted illustrations in black and blue reflect the boys' mischievous antics and capture the humor of each noise-filled activity. Occasional glimpses are seen of the parents in their vain attempts to sleep. In final desperation they run downstairs to find that it is 9:45 exactly! The passage of time is reinforced with the appropriate placement of a clock face within the illustrations. The times depicted require a full understanding of time telling to estimate the minutes to and after the hour. Originally published in 1968, *Sunday Morning* continues to delight readers.

Measurement of Time— Concepts

Axelrod's *Pigs on a Blanket* was the choice of Carrie Eipperle, the student-teacher in Chris Hartman's classroom. Prior to reading the book to the second graders, Carrie discussed half and quarter hours to help the children understand the language in the book, such as quarter to and half past. Circular fraction pieces and fraction rings that have numerals printed on the circle's circumference to indicate five-minute intervals helped to illustrate these ideas. As Carrie began to read the story, a Judy clock was used to record the passing times. The children anticipated with delight the antics of the Pig family, and as they heard that more time had passed, they took turns recording the changes. The next day, each student had his or her own clock and indicated each time as Carrie read the book a second time. The second graders then wrote their own beach stories using elapsed time. After trading stories and figuring out the times, the children compiled the stories into a class book.

Moments, minutes, days, weeks, months, and years are concepts with which primary-aged children are still coming to terms. The books annotated in this section of the bibliography develop these concepts of time. Hutchins's *Clocks and More Clocks* is a humorous introduction to time concepts and reading clocks; Taber's *The Boy Who Stopped Time* dramatically presents another aspect. Axelrod's delightful *Pigs on a Blanket* combines the passage of time with time measures. The passage of time from sunrise to sunset is captured in Gerstein's *The Sun's Day;* Carle and Ward investigate longer intervals, such as the passage of a week. Halsey, Lillie, Singer, and Sendak illustrate time passages over a year. Reiser's *Christmas Counting* looks at changes over the passage of years. Activities to develop time concepts abound in both Llewellyn's *My First Book of Time* and Burns's *This Book Is about Time*. Experiments on time can also be found in the books by Smoothey and Walpole.

The following book are out of print but are considered exceptional:

Coats, Laura Jane. *The Oak Tree*

Gould, Deborah. *Brendan's Best-Timed Birthday*

Kherdian, David, and **Nonny Hogrogian.** *Right Now*

Krensky, Stephen. *Big Time Bears*

Maestro, Betsy, and **Giulio Maestro.** *Through the Year with Harriet*

Neasi, Barbara J. *A Minute Is a Minute*

Copies may be available in your library. See the first edition of *The Wonderful World of Mathematics* for annotations.

Anno, Mitsumasa. *Anno's Counting Book.* New York: Thomas Y. Crowell Publishers, 1977.

See Early Number Concepts: Counting—Zero

Axelrod, Amy. *Pigs on a Blanket.* Illustrated by Sharon McGinley-Nally. New York: Simon & Schuster Books for Young Readers, 1996.

ISBN 0-689-80505-5 $13.00

★★★ Single concept, PS–3

Like Axelrod's *Pigs Will Be Pigs,* this humorous narrative exploring the passage of time stars Mr. and Mrs. Pig and their two piglets. One hot day they decide to break out of their rut and head to the beach, which is an hour away. Since the decision was not made until 11:30, they hurry to get ready. In ten minutes the piglets are ready to leave, and in five more minutes Mrs. Pig is also ready. But they wait and wait for forty-five minutes for Mr. Pig, who could not find a swimsuit that fit. At 12:45 they are in the car and ready to go, but an hour passes before Mr. Pig finally finds the car keys. The story line continues as new delays occur: twenty-five minutes for the train to pass over the crossing or thirteen minutes when Mr. Pig receives a speeding ticket. Just as the Pig family is finally ready to swim, the lifeguard announces the 5:30 closing of the beach. "How did the Pigs run out of time?" is the question posed to the reader. Throughout the story passages of time are given through clock faces or a statement on the length of time for different events. Readers can trace the passage of time as they read the story as well as the summary at the end of the book. Originally composed in inks, watercolors, and acrylics, McGinley-Nally's riotously funny cartoon illustrations are dominated by bright colors and wild prints. From Mr. Pig's black-and-white striped swimsuit and matching sunglasses to the alligator lifeguard with his safari hat, the details in each scene add to the zany humor.

Aylesworth, Jim. *One Crow: A Counting Rhyme.* Illustrated by Ruth Young. Philadelphia: J. B. Lippincott Co., 1988.

See Early Number Concepts: Counting—Zero

Burns, Marilyn. *This Book Is about Time.* Illustrated by Martha Weston. Boston: Little, Brown & Co., 1978.

ISBN 0-316-11752-8 $15.95
ISBN 0-316-11750-1   (paper)   11.95

★★★ Multiconcept, 5+

Historically, yellow, orange, red, purple, and green were the names the people of the Soviet Union gave to the days in their five-day week—this is one of the many intriguing historical facts interwoven with the information and activities in this book about time. The ten chapters can be read one section at a time and are cross-referenced for some activities. Throughout the book, the readers are asked to interview others or to look at their personal uses of time to find whether time is noted by intervals or events or is measured by clocks or calendars. How people used time in the past is linked to their measurement of time. The historical and cultural facts surrounding the origins of daylight saving time, the International Date Line, and the time zones in the United States are given. Different inventions for measuring the day are described, and the activities include making and experimenting with sundials, water clocks, sand clocks, and clocks with pendulums. A later activity suggests observing and collecting data on the moon's cycles for a month. Estimation is included; the reader is asked to predict the length of events that would occur in seconds or minutes. The reader is invited to participate in activities to develop these concepts or is involved in designing experiments to test the estimates. The end of the book considers longer intervals. A fascinating description of a one-year-long movie about our universe from its origin to the present is included. Perceptions of time are extended to scientific descriptions of jet lag and the internal clocks of plants, animals, and humans.

Carle, Eric. *Today Is Monday.* New York: Philomel Books, 1993.

ISBN 0-399-21966-8 $14.95

☆ Single concept, PS–2

Carle's colorful collages introduce a new food and each day of the week starting with Monday. For Thursday, orange, gray, and brown are the predominant colors used to create a green-eyed cat eating roast beef. Each new layout depicts a different animal and summarizes the previous pages: Thursday, roast beef/Wednesday, ZOOOOP/Tuesday, spaghetti/Monday, string beans. A macaw invites all hungry children to come and eat. Children of different nationalities, with one child in a wheelchair, are depicted around a dining room table. On the wall behind them are smaller versions of the animals illustrated in the book. A final page contains a musical score and the verses reflecting the story.

————. *The Very Hungry Caterpillar.* New York: Philomel Books, 1987.
ISBN 0-399-20853-4 $6.95
ISBN 0-399-22753-9   (library binding)   17.95
ISBN 0-399-21301-5   (Putnam, 1986)   4.95
★★ Single concept, K–2

A hungry caterpillar emerges from an egg on Sunday. The next pages show the one apple he ate on Monday, the two pears he ate on Tuesday, and so on. Each fruit on the back and front of a page has a hole through it. In stair-step fashion, each page from 1 through 5 is longer so that the reader can see, for instance, that three plums are more than two pears. On Saturday, the caterpillar eats ten different foods before he spins his cocoon. Two weeks later he emerges as a colorful butterfly, shown on the last page. Brilliant collages effectively illustrate the story. More than 4 million copies of this classic, which was originally published in 1969, have been sold.

Cassidy, John. *The Time Book.* Illustrated by Roger Bollen. Palo Alto, Calif.: Klutz Press, 1991.
See Measurement: Measurement of Time—Telling Time

Florian, Douglas. *A Summer Day.* New York: Greenwillow Books, 1988.
ISBN 0-688-07564-9 $11.95
ISBN 0-688-07565-7 (library binding)   11.88
★★ Single concept, PS

The passage of time is clearly and simply illustrated in this charming rhyme, which takes the reader from sunrise to sunset on a summer day as a family takes a trip to the countryside. The changes in morning and evening light are captured by the pleasing illustrations in crayon and ink.

Gerstein, Mordicai. *The Sun's Day.* New York: Harper & Row Publishers, 1989.
ISBN 0-06-022404-5 $13.00
ISBN 0-06-022405-3   (library binding)   14.89
★★ Single concept, PS–1

Abstract pastel illustrations dramatically present each scene from sunrise to sunset. The day progresses hour by hour, with each busy scene detailing the activities for that hour. The sun's daylight journey is the focus of these layouts, and the sun in each symbolizes a new entity in surprising forms. In one of the first scenes, clouds in the shapes of a hen and a rooster hover over the sun, which appears as a chick popping out of an egg; the landscape below shows a farmstead starting to come to life at dawn. A scene at two o'clock shows the skyline of San

Francisco, with a peach orchard across the bay; the sun appears as a ripe, luscious peach. The fascinating illustrations should capture the interest of readers and could promote discussions of these scenes and of what happens in the readers' lives during each time period.

Halsey, Megan. *Jump for Joy: A Book of Months.* New York: Bradbury Press, 1994.

ISBN 0-02-742040-X $ 14.95

☆ Single concept, PS–1

Alliterative directions provide an active and satisfying introduction to the calendar year: "Jump for Joy in January," "Follow a Fox in February," ..., "Do a Dance in December." Changes of the seasons; celebrations of different events, such as New Year's and the beginning of the school year; and activities appropriate to a given month are presented throughout *Jump for Joy.* The accompanying illustrations reflect the action of the text through their vibrant watercolors, dyes, and pen and ink. The interesting three-dimensional effects of the paper images, which were cut out, sculpted, and assembled, may inspire young artists to try their hand at similar techniques. The last page calls for a recital and review of the twelve months' names.

Hartmann, Wendy. *One Sun Rises: An African Wildlife Counting Book.* *See* Early Number Concepts: Number in Other Cultures

Hutchins, Pat. *Clocks and More Clocks.* New York: Aladdin Books, 1994.

ISBN 0-689-71769-5 $4.95

★★ Single concept, PS-3

Originally published in 1970, this humorous book about the passage of time was reissued in paper by Aladdin Books. The dilemma begins when Mr. Higgins questions whether the clock in his attic is correct. After buying a new clock and placing it in his bedroom, he finds that the times are not the same. He continues buying clocks for different areas of his house, but as he checks the time he always finds that they are not the same. The clock maker solves the problem by comparing each clock's time with that on a wristwatch. This story provides a humorous introduction to the movements of a clock and the passage of time, since Mr. Higgins does not relate the two ideas. The appealing cartoon illustrations were created with pen-and-ink lines and watercolors.

Lewis, J. Patrick. *July Is a Mad Mosquito.* Illustrated by Melanie W. Hall. New York: Atheneum, 1994.

ISBN 0-689-31813-8 $14.95

☆ Single concept, PS–2

This collection of poems captures the sights, sounds, smells, and mood of each month of life in the United States. From the refrigerated paradise of January to the diamonds of frost in November to Christmas goose in December, each poem provides rich detail. The poems are of varying quality with the last being the finest; this lyrical poem weaves the major points from each of the preceding twelve to pass through the year again. The illustrations that accompany the text have a dreamlike quality appropriate to the context. Seasonal changes are reinforced by the illustrations, which depict appropriate weather, use well-chosen paint colors, and show the characters dressed in seasonal clothing and pursuing appropriate activities.

Lillie, Patricia. *When This Box Is Full.* Illustrated by Donald Crews. New York: Greenwillow Books, 1993.

ISBN 0-688-12016-4 $14.00

ISBN 0-688-12017-2 (library binding) 13.93

★★ Single concept, PS–2

A thoughtful little African American child studies an empty box and proceeds to tell what she will fill it with and, finally, for what purpose. The opening page simply shows the empty wooden box. At the top of the next page appears the word *January* with a picture of a woolen scarf being added to the box and "snowman's scarf" written beneath. As the months of February and March appear on the list, a red valentine heart and a robin's feather are placed in the box. As the year progresses, items that represent each month are added. In December, a silver star is added and the box is full. And then the little girl makes an offer: "I will share it with you." An African American boy is shown sitting beside the full box, his mouth open with amazement. Minimal images from black-and-white photographs that were hand-colored with pastels show the gradually filling wooden box. This unique book provides an alternative to the many calendar books available.

Llewellyn, Claire. *My First Book of Time.* New York: Dorling Kindersley, 1992.

ISBN 1-879431-78-5 $14.95

★★ Multiconcept, PS–2

Both the passage of time and learning to tell time are introduced in this oversize primary book. Watercolors are effectively combined with color photographs to illustrate children of different ethnic groups actively exploring time concepts. Children are shown engaged in various activities to illustrate day and night; morning, afternoon, and evening; and longer blocks of time, such as the days of the week and the four seasons. An explanation and diagram of the earth's rotation and its rotation about the sun are used to explain night and day and the changing seasons. Change over time is also explored through different stages of growth with illustrations depicting changes from bean seeds to bean sprouts in six days, a duckling to a duck in eight weeks, a baby to a mature woman in fifty-eight years. An eclectic collection of clocks and watches with descriptions of their uses are a bridge to the section on telling time. Twelve clock faces show each hour from one o'clock through twelve o'clock; half and quarter hours are effectively shown by shading in the appropriate quarter, half, or three-quarters area of the clock face. Telling time by five-minute intervals includes both minutes after and minutes to the hour; twenty-four-hour clocks are also introduced. All these activities relate telling time on analog clocks with telling time on digital clocks. To connect the concept of time with telling time, activities are described that involve quarter hours and seconds. Informal and past representations of telling time include such examples as sand timers, sundials, and candles. Questions are posed throughout the book; some are to help connect time to the reader's world; others are to check the reader's understanding of the concepts presented in the book. One activity includes matching pairs in a six-by-seven grid that contains digital and analog clocks, twenty-four-hour clocks, and word names for telling time; another activity includes directions for building a calendar. An end page contains a clock face with a red plastic hour hand and a blue minute hand; this page can be folded out and used with any page throughout the book. Other special features include a table of contents, a note to parents and teachers, and a glossary. Because of the vast amount of information, this book should appeal to a variety of readers.

Pluckrose, Henry. *Time.* Illustrated by Chris Fairclough. Chicago: Children's Press, 1995.

See Series and Other Resources: Math Counts

Powers, Christine. *My Day with Numbers.* New York: Scholastic, 1992.

See Early Number Concepts: Counting—One to Ten

Reiser, Lynn. *Christmas Counting.* New York: Greenwillow Books, 1992.

See Early Number Concepts: Counting—One to Ten

Rothman, Joel. *A Moment in Time.* Illustrated by Don Leake. New York: Scroll Press, 1973.

ISBN 0-87592-034-9 $7.95

☆ Single concept, PS–2

Colorful paintings with broad brush strokes and very few words capture a moment in time—an apple falling in slow motion. This pleasing book graphically represents sequence and the concept of time. The idea of such a fleeting moment might be very difficult to understand for the preschool and primary children for whom the book was designed. Since these concepts need to be experienced, this book could be used to promote a discussion and activities about time. The boy's picking up and eating the apple seems to be a completely separate event that destroys the attempt to capture "a moment in time."

Russo, Marisabina. *Only Six More Days Left.* New York: Greenwillow Books, 1988.

ISBN 0-688-07071-X		$11.95
ISBN 0-688-07072-8	(library binding)	11.88
ISBN 0-14-054473-9	(paper, Puffin Books)	3.99

★★ Single concept, PS–3

Ben gleefully counts down the six days to his birthday and constantly talks about the impending event. Through each successive day, the text and expressive illustrations clearly show that Ben's older sister Molly is tired of the topic and does not want to take part in the celebration. The problem is tactfully resolved, and Molly joins her brother and his friends at the party. Children from various cultures are included in the attractive illustrations painted in gouache. Although one of this book's main purposes is to explore sibling relationships, it can also be used to generate countdowns to other celebrations. Molly's letter at the end of the book might motivate a primary class to do some letter writing about time experiences.

Sendak, Maurice. *Chicken Soup with Rice: A Book of Months.* New York: HarperCollins Children's Books, 1962.

ISBN 0-06-025538-8	(library binding)	$13.89
ISBN 0-06-443253-X	(paper)	3.95
ISBN 0-590-63088-1	(Scholastic)	5.95
ISBN 0-590-41033-4	(paper, Scholastic)	2.50

★★ Single concept, K–3

One of Sendak's characteristically rambunctious little boys takes the reader through the months of the year while eating, serving, or playing in chicken soup with rice. The story is usually realistic in describing activities for each month; however, some months are pure fantasy. Simple drawings complement the story line. The rhyme is compelling and the pattern predictable; the words change enough to keep the reader concentrating. Young listeners and readers catch on to the rhythm quickly and enjoy "singing" the story. This is a good story for children to rewrite using their own examples for each month.

Singer, Marilyn. *Turtle in July.* Illustrated by Jerry Pinkney. New York: Macmillan Publishing Co., 1989.

| ISBN 0-02-782881-6 | $13.95 |

★★ Single concept, PS-3

Pinkney's magnificent watercolors and Singer's evocative verse are a superb partnership in creating sensory images month by month as the seasons change. Each of the four selections reflects the changing seasons through illustrations and verse: "in autumn / I settle / belly down in the shallows / above me / leaves / red and yellow / spinning slowing / in the wind and water.... " From the doleur of the bullheads in icy waters of winter, the swiftness of the deer running over the hard-packed snow of January, the search and sweep of the barn owl through night skies in February, the clumsy confusion of the waking bear in March, or the escape of the turtle from the heat in July, the images are carefully crafted for the lucky reader.

Smoothey, Marion. *Time, Distance, and Speed.* Illustrated by Ted Evans. New York: Marshall Cavendish, 1993.

See Series and Other Resources: Let's Investigate

Taber, Anthony. *The Boy Who Stopped Time.* New York: Margaret K. McElderry Books, 1993.

ISBN 0-689-50460-8 $13.95

★★★ Single concept, PS–3

Elegant black-and-white pencil drawings illustrate this delightful fantasy of the boy who stopped time. Dreading his 7:30 bedtime, Julian asks his mother if he can stay up later to watch television. She denies his request and tells him to be in bed before the minute hand on the clock reaches the 6. After she leaves to tuck in his little sister, Julian decides to stop the clock's pendulum. Immediately a hush falls over the home, and Julian discovers that all activities surrounding him have stopped: the television screen has frozen on one scene; his mother is leaning over the crib like a statue; his father is standing where he was tossing rocks on a pile, with a rock suspended in midair. Julian continues to explore the farm and the neighboring town and finds all the animals and people suspended in time. The silence leads to a feeling of sadness and a wish to return to normal, so Julian returns and starts the clock. This delightful tale has stunning illustrations that are effective in depicting the eerie sensation of silence and stillness that results from stopping the passage of time.

Time-Life for Children Staff. *Alice in Numberland: Fantasy Math.* Alexandria, Va.: Time-Life for Children, 1993.

See Series and Other Resources: I Love Math

———. *The Case of the Missing Zebra Stripes: Zoo Math.* Alexandria, Va.: Time-Life for Children, 1992.

See Series and Other Resources: I Love Math

———. *From Head to Toe: Body Math.* Alexandria, Va.: Time-Life for Children, 1993.

See Series and Other Resources: I Love Math

———. *Play Ball: Sports Math.* Alexandria, Va.: Time-Life for Children, 1993.

See Series and Other Resources: I Love Math

———. *Right in Your Own Backyard: Nature Math.* Alexandria, Va.: Time-Life for Children, 1993.
See Series and Other Resources: I Love Math

———. *The Search for the Mystery Planet: Space Math.* Alexandria, Va.: Time-Life for Children, 1993.
See Series and Other Resources: I Love Math

———. *See You Later Escalator: Mall Math.* Alexandria, Va.: Time-Life for Children, 1993.
See Series and Other Resources: I Love Math

Ward, Cindy. *Cookie's Week.* Illustrated by Tomie dePaola. New York: G. P. Putnam's Sons, 1988.
ISBN 0-399-21498-4 $10.95
☆ Single concept, PS–1

On Monday the mischievous black-and-white cat Cookie falls into the toilet; on Tuesday Cookie knocks over flowerpots. The days of the week continue sequentially with a new disaster each day. The story ends by contemplating tomorrow, which is Sunday, and hoping that Cookie will rest. The simple text, complemented by pastel illustrations, could promote a discussion on the reader's life on different days of the week.

Measurement of Time— Extensions and Connections

The origins of time measurement connect history, science, and mathematics and provide the theme for most of these books. Directions for making many wonderful timing devices of the past are found in works by Branley, Smoothey, and Walpole. The history of the calendar is thoroughly presented in the work of Branley. Gerstein's *The Story of May* is a playful presentation of the story of naming the months of the year. Time zones around the world are perceptively represented by Baumann in *What Time Is It around the World?* and by Singer in *Nine O'clock Lullaby.*

The following books are out of print but are considered exceptional:

Anno, Mitsumasa. *Anno's Sundial*

Apfel, Necia H. *Calendars*

Jennings, Terry. *Time*

Brindze, Ruth. *Story of Our Calendar*

Copies may be available in your library. See the first edition of *The Wonderful World of Mathematics* for annotations.

Baumann, Hans. *What Time Is It around the World?* Illustrated by Antoni Boratynski. New York: Scroll Press, 1979.

ISBN 0-87592-061-6 $6.95

★★ Single concept, K–5

Beautiful, bright, eye-catching illustrations graphically show the time of day in various parts of the world as the reader progresses through twenty-four time zones. The point of the book—that when it is 7:00 A.M. in one place, it is progressively an hour later in other places in the world as one moves east—is not clearly expressed. Some explanation would help the reader understand that at the same time it is 7:00 A.M. in London, it is 8:00 A.M. in Morocco, 9:00 A.M. in the Middle East, and so on. With such adult guidance, this book is recommended to help children understand time around the world.

Branley, Franklyn Mansfield. *Keeping Time: From the Beginning and into the 21st Century.* Illustrated by Jill Weber. Boston: Houghton Mifflin, 1993.

ISBN 0-395-47777-8 $13.95

☆ Single concept, (3–6)

Although the sun, moon, and stars were the timekeepers for early people, today's need for schedules in which minutes and even seconds become important requires everyone to use clocks. This book explores how people learned to keep track of time. Directions are given for making a sundial, candle clock, water clock, and sand clock. All were used before the first mechanical clock was developed in Europe in the thirteenth century. An increasing need for accuracy in keeping time has led to the development of the digital clock and the atomic clock. Accuracy can be measured in micro-, nano-, pico-, and femtoseconds. Other topics include a discussion of what time is; standard and daylight time; time zones; the reasons for twenty-four hours in a day, sixty minutes in an hour, sixty seconds in a minute, seven days in a week, and twelve months in a year; whether days are getting longer; and when or where the day begins. The Gregorian calendar is explained and finally time is discussed in relation to space and the potential for colonizing other planets. The text is simple, direct, and clearly stated. Interesting and amusing facts coupled with humorous line drawings should prove motivating to the reader. Diagrams and charts are of varying quality and usefulness for clarifying the text; one questionable example is the directions for making a sundial. The table of contents lists seventeen chapters in this 105-page book; an index and a bibliography of other books on time for young readers and one for more advanced readers are also included.

Burns, Marilyn. *This Book Is about Time.* Illustrated by Martha Weston. Boston: Little, Brown & Co., 1978.

See Measurement: Measurement of Time—Concepts

Fisher, Leonard Everett. *Calendar Art: Thirteen Days, Weeks, Months, Years from around the World.* New York: Four Winds Press, 1987.

ISBN 0-02-735350-8 $15.95

☆ Single concept, 4–6

After a brief introduction, which explores the need for the accurate recording of time, the author presents the various ways cultures throughout history have responded to this need. Calendars of many cultures and regions are presented and discussed: Aztec, Babylonian, Chinese, Egyptian, French Revolutionary, Gregorian, Hebrew, Islamic, Julian, Mayan, Roman, and Stonehenge. Each presentation includes a history of the calendar, how the calendar works, and artwork. The world calendar also appears in this interesting history- and mathematics-related book.

Gerstein, Mordicai. *The Story of May.* New York: HarperCollins, 1993.
ISBN 0-06-022289-1 $15.89
☆ Single concept, PS–3

Gerstein creates a fantasy about the months of the year through both the story line and the elaborate watercolor illustrations. The tale starts as April teaches her daughter May her responsibilities for her month of May, such as scattering wildflowers and welcoming returning birds. As May wanders through the day, she travels too far and encounters warmer temperatures, greener trees, and her Aunt June. Her travels continue from one relative to the next, each of whom lives in and personifies the next month. The cyclical notion of the calendar is noted by May's journey through January, February, and March to return to her mother after a visit with her father, December. Creative "explanations" for the order of the months and for why May breezes can sometimes occur on a winter day are encountered in the book.

Massin. *Fun with 9umbers.* Illustrated by Les Chats Pelés. San Diego, Calif.: Creative Editions, 1993.
See Number—Extensions and Connections: Number Systems

Morgan, Rowland. *In the Next Three Seconds.* Illustrated by Rod Josey and Kira Josey. New York: Lodestar Books, 1997.
See Number—Extensions and Connections: Statistics and Probability

Singer, Marilyn. *Nine O'clock Lullaby.* Illustrated by Frane' Lessac. Harper Collins Publishers, 1991.
ISBN 0-06-025648-6   (library binding) $14.89
ISBN 0-06-443319-6   4.95
★★ Single concept, PS–3 (3–5)

Take a trip around the world with Marilyn Singer as different time zones are visited at the same hour. Each colorful, childlike illustration captures appropriate activities and scenes for the hour and the culture visited at each time zone location. The opening scene depicts 9 P.M. traffic in Brooklyn, New York, while inside an apartment building, a mother reads a bedtime story to her child. At the same time, it is 10 P.M. in Puerto Rico; the second illustration shows people dancing at a special outdoor party where sweet rice, fruit ice, and coconut candy are being served. Other locations include 10 A.M. in Guangzhou, China, where busy bicycle riders pedal past a market, and 3 P.M. in Samoa, where adults are weaving grass mats in a guest house and children are playing along the ocean shore. Our trip around the world is completed as we return to 9 P.M. traffic in Brooklyn. The author's notes at the end of the book briefly discuss time changes as the earth rotates on its axis. The creation of time zones in 1884 helped alleviate confusion. Also included is an explanation of the half-hour time depicted in the 7:30 A.M. visit to

India, where women are gathering water at a well. The author's notes state that India's time zone is one-half hour behind the time zone to the east and one-half hour ahead of the time zone to the west.

Smoothey, Marion. *Time, Distance, and Speed.* Illustrated by Ted Evans. New York: Marshall Cavendish, 1993.

See Series and Other Resources: Let's Investigate

Walpole, Brenda. *Time.* Illustrated by Dennis Tinkler and Chris Fairclough. Milwaukee: Gareth Stevens, 1995.

See Series and Other Resources: Measure Up with Science

Money

Younger readers will delight in the realistic illustrations of coins in Hoban's *Twenty-six Letters and Ninety-nine Cents;* values of money and equivalent values among coins are explored in works by Hoban and McMillan. Axelrod, Viorst, and Zimelman effectively use humor to engage readers in situations involving money. Schwartz's fanciful examples depicting what can be bought for various sums of money and illustrating these values in graphic terms are captivating and help develop number sense. Money's connection to history and to other cultures is the main theme of Mitgutch's *From Gold to Money.*

Betsy and Giulio Maestro's *Dollars and Cents for Harriet* is out of print but is considered exceptional; a copy may be available in your library. See the first edition of *The Wonderful World of Mathematics* for an annotation.

Axelrod, Amy. *Pigs Will Be Pigs.* Illustrated by Sharon McGinley-Nally. New York: Simon & Schuster Books for Young Readers, 1994.

ISBN 0-02-765415-X $15.00

★★ Single concept, K–3

Bright colors and wild busy prints dominate the illustrations of the Pig family and their zany antics as they search their home for enough money to buy dinner at their favorite restaurant. As they hunt for enough money, Mrs. Pig finds two nickels, five pennies, and one quarter to add to the two-dollar bill that Mr. Pig had hidden in his socks. The piglets found six dimes in the toy chest, a dollar bill on the book shelf, and two hundred pennies in their penny collection. After finding more money, they set off for the Enchanted Enchilada. Complete with appetizers, soups, salads, desserts, beverages, and specialties, the menu that the waitress hands them is reproduced on a separate page in the book. The questions posed at the book's end involve how much money the Pigs found, spent, and had left. The reader is also encouraged to choose other menu items that would cost the same amount of money.

Brown, Marc Tolon. *Arthur's TV Trouble.* Boston: Little, Brown & Co., 1995.

ISBN 0-316-10919-3 $14.95

Single concept, PS–3

Problem-solving situations involving money are part of the plot in *Arthur's TV Trouble,* by popular author and illustrator Marc Brown. When Arthur sees advertisements for the amazing Treat Timer for dogs, he decides he must buy one for his dog, Pal. But the Treat Timer costs $19.95 and Arthur has only $10.03. The story line reflects Arthur's concerns as he talks with his parents about advancing him money and his distractions as he ponders how to get the extra money. Mr. Sipple offers to pay him fifty cents a stack to recycle all the newspapers stored in his garage. After a couple of mishaps, Arthur manages to bundle all twenty-four stacks of paper and collects twelve dollars. In addition to earning money, the story line involves Arthur's disappointment with the Treat Timer because it looks smaller than shown on television, takes five hours to assemble, and is promptly ignored by Pal. Brown's charming illustrations depict Arthur, his family, and friends in settings familar to the young reader.

Hoban, Tana. *Twenty-six Letters and Ninety-nine Cents.* New York: Greenwillow Books, 1987.

ISBN 0-688-06361-6		$15.00
ISBN 0-688-06362-4	(library binding)	14.93
ISBN 0-688-14389-X	(paper, Mulberry)	4.95

★★★ Single concept, PS–3

Glossy, brightly colored numerals were photographed with the corresponding value of coins. The values from 1 through 30, 35, 40, 45, 50, 60, 70, 80, 90, and 99 are represented. Some values are shown in more than one way—for example, 6 is six pennies or one nickel and one penny; 10 is ten pennies, two nickels, or one dime. Usually the least possible number of coins is shown—17 is one dime, one nickel, and two pennies; 80 is three quarters and one nickel. The book could easily lead the reader to explore other ways of showing the same value. Note: After reaching 99, the reader can turn the book around to explore the twenty-six letters of the alphabet (lowercase and uppercase) with a corresponding pictured object (e.g., Ww and a picture of a wagon).

Leedy, Loreen. *The Monster Money Book.* New York: Holiday House, 1992.

ISBN 0-8234-0922-8 $14.95

☆ Single concept, 1–4

Sarah's and Grub's introduction to the Monster Club has far-reaching results. Grub must work for money to pay dues. At the introductory meeting, they find themselves involved in a debate on how to spend the fifty-four dollars in the club treasury. Dialogues between the monsters and Sarah, the lone human—and potential club member—include diverse comments on how to earn money as well as how to spend it. A budget is proposed that includes categories of saving, spending, and giving. From this discussion, simple explanations of consumer concepts emerge, such as investments, profit, interest, withdrawals, checks, and bank cards. Sarah guides the discussion by explaining many of these things and by describing how to be a smart shopper. Explanations are supported by subtly colored pastel drawings, clear diagrams, and charts. All text is written as conversations with speech boxes above the characters' heads. The fact that all but one of the characters are monsters is irrelevant to the story and is only supported by weird actions like those of Grub, who eats pickle jelly sandwiches and drinks turnip juice. A glossary at the back of the book reinforces an understanding of the most important terms.

McMillan, Bruce. *Jelly Beans for Sale.* New York: Scholastic Press, 1996.

ISBN 0-590-86584-6 $15.95

★★★ Single concept, PS–3

"One for a penny. / Ten for a dime. / Count them and buy them. / You'll have a good time!" Alongside a piggy bank and against a background of shiny, brightly colored jelly beans, this invitation is posed on an opening page. A subsequent chart lists the values for a penny, a nickel, a dime, and a quarter with photographs of both sides of each coin. Each left-hand page shows a close-up of a purchase of jelly beans, for example, one nickel for five jelly beans or one dime, two nickels, and five pennies for one group of ten jelly beans, two groups of five yellow jelly beans, and five white jelly beans. The respective captions read 5¢ = 5 jelly beans and 1¢ + 1¢ + 1¢ + 1¢ + 1¢ + 5¢ + 5¢ +10¢ = 25 jelly beans. Each right-hand page shows one or more children with jelly beans of the same color. Each pair of pictures is bordered on a larger photograph that is a close-up of lots of jelly beans; the pictures and the borders are color coordinated with the pictured jelly beans. For example, when five yellow and five red jelly beans are purchased with a nickel and five pennies, the child shown on the opposite page is dressed in bright yellow and red; the borders for both pages are close-ups of yellow and red jelly beans. Various combinations of coins and groupings of jelly beans are used throughout the book. Some of the representations for twenty-five cents include different combinations of nickels, pennies, and dimes as well as one quarter; the jelly beans are shown in groupings of fifteen and ten; five, five, five, and ten; twenty and five; and 5 fives. McMillan's photographs of children from various ethnic backgrounds buying and enjoying jelly beans are a delight; a final photograph shows children placing coins in the piggy bank. This is an excellent introduction to values of coins and their interrelationships as well as to different ways to partition numbers.

Mitgutch, Ali. *From Gold to Money.* Minneapolis: Carolrhoda Books, 1985.

ISBN 0-87614-230-7 $15.95

☆ Single concept, K–3

The progression from bartering to using gold, coins, and paper money is the focus of this "start to finish" book. The limitations of each system and why the changes occurred are described. Brightly colored and framed illustrations show people trading or buying materials and services in a variety of ways. This book is a basic introduction to the history of money.

Schwartz, David M. *If You Made a Million.* Illustrated by Steven Kellogg. New York: Lothrop, Lee & Shepherd Books, 1989.

ISBN 0-688-07017-5 $16.00
ISBN 0-688-07018-3 (library binding)   15.93
ISBN 0-688-13634-6 (paper, Morrow)   4.95

★★★ Single concept, 1–5

Similar to the format of *How Much Is a Million?*, Schwartz and Kellogg's fantasy examples and detailed, pastel drawings cleverly introduce money from 1 penny to 1 million dollars. The examples illustrated by Marvelosissimo, the Mathematical Magician and his eight friends include tasks to earn a particular amount of money, such as feed the fish to earn 1 penny or bake a cake to earn $5. Also, the examples include items to buy, such as a hippopotamus for $1000 or tickets to the moon for $1 000 000. Photographs of coins and bills show relationships; 1 dollar bill equals 4 quarters, 10 dimes, 20 nickels, or 100 pennies. As the denominations get larger, comparisons are used; a fifty-foot stack of pennies would be $100; a school bus filled with nickels would be $1 000 000. Marvelosissimo and his friends also become involved with writing checks and earning interest on money deposited in a bank. Three pages of notes from the author explain how he generated the data in the examples. These notes also explain such concepts as monetary systems, banks, interest and compound interest, checks and checking accounts, loans, and income tax. All the notes are extensions of the excellent examples throughout the book.

Time-Life for Children Staff. *See You Later Escalator: Mall Math.* Alexandria, Va.: Time-Life for Children, 1993.
See Series and Other Resources: I Love Math

Viorst, Judith. *Alexander Who Used to Be Rich Last Sunday.* Illustrated by Ray Cruz. New York: Atheneum Publishers, 1979.

ISBN 0-689-30602-4   (library binding) $13.95
ISBN 0-689-71199-9   (paper, Aladdin)   3.95

★★ Single concept, K–2

Alexander believes, "It isn't fair that my brother Anthony has two dollars and three quarters and one dime and seven nickels and eighteen pennies," but Alexander has only bus tokens. The story line and the pen-and-ink drawings effectively illustrate the humorous plight of Alexander, who used to be rich last Sunday when he received a dollar from his grandparents. His money soon disappears as he pays a ten-cent fine for swearing at his brother, rents Eddie's pet snake for an hour for twelve cents,   wagers and loses fifteen cents. It is left for the reader to determine whether Alexander has lost all his money by accounting for all his expenses.

Wells, Rosemary. *Bunny Money.* New York: Dial Books for Young Readers, 1997
ISBN 0-8037-2146-3
ISBN 0-8037-2147-1   (library binding)
Single concept, PS–3

Rosemary Wells has another winning book with Max and Ruby's adventures in *Bunny Money.* An opening page shows the contents of Ruby's red wallet—1 five-dollar bill and 10 one-dollar bills—as the children set off to buy Grandma's birthday present. The first dollar pays for bus fare; two more dollars is spent for vampire teeth filled with cherry syrup. More of the dollars disappear as emergencies and other distractions arise. After spending their last bill on presents for Grandmother, Ruby and Max realize that no money is left for the bus fare home. Max produces his lucky quarter, which is used to telephone Grandmother to request a ride home. Well's pleasing illustrations and the humorous twist to the story's ending will delight young readers. "Bunny money" to be photocopied is pictured on the inside covers of the book so readers can act out the story or create their own stories. An added bonus to promote further discussion is the "portraits" on the bunny money, which depict such famous people as Jessye Norman, Martina Navratilova, Desmond Tutu, Yo-Yo Ma, Jonas Salk, and Frida Kahlo.

Zimelman, Nathan. *How the Second Grade Got $8205.50 to Visit the Statue of Liberty.* Illustrated by Bill Slavin. Morton Grove, Ill.: Albert Whitin & Co., 1992.
ISBN 0-8075-3431-5 $12.95
★★★ Single concept, K–3 (2–5)

Join "Susan Olson, treasurer and reporter, second grade, Newton Barnaby School" as she writes her report on the many creative and hilarious fund-raising projects the children engage in as they attempt to make money for a trip to the Statue of Liberty. Although most of the fund-raising events are realistic, there are twists to the stories that will delight and surprise the reader. The tongue-in-cheek text, which gives sometimes painfully accurate, but always humorous, insights into adult behavior, is accompanied by whimsical watercolors that capture the expressions of the characters and the moods of the moment. This thoroughly enjoyable story can be experienced at many levels of understanding. Each episode of Susan's report starts with *expenses,* describes the situation, and ends with *profits.* Readers can keep track of how much money the second graders are accumulating; at times they have a negative balance. Zimelman and Slavin present an excellent opportunity for reader involvement, which should lead to several purposeful rereadings of the text. Additionally, readers can use it as a springboard for writing fund-raising stories of their own.

Geometry and
Spatial Sense

W HAT kind of pie do you like best? Apple? Cherry? Huckleberry? Banana cream? When you cut a piece of pie, you cut an angle. One way to describe an angle is to say that it is the opening between two lines that meet each other." The thirty-six third graders from a class in rural southern Illinois listened carefully as their teacher read Froman's *Angles Are Easy as Pie* to them. One of the third graders drew a diagram on the chalkboard showing a pie cut into four pieces. Another child drew a pie cut into six pieces. The children decided that if they were hungry and if it was their favorite pie, they would choose a piece from the pie cut into four pieces. Another of the book's activities involved them in exploring the angles of a triangle. After carefully labeling each angle of a triangle with a number and shading each angle, the children tore the angles from the triangle and arranged them to find that the three angles formed half a pie.

Each semester in their mathematics class, preservice teachers at the University of Northern Iowa participate in the same activity from this classic. Like the third graders, they become involved and express delight with the book's characters and activities. The college students extend the ideas to explore quadrangles and pentagons and to form a generalization about how to determine the number of degrees for a polygon with n sides. Not only are they starting with an intuitive, concrete approach to form their own conjectures, they are also becoming acquainted with activities appropriate for elementary school children.

Although Froman's *Angles Are Easy as Pie* is out of print, it exemplifies one of the many quality children's books that contain activities in geometry. Books that relate directly to geometry are divided into three groups. Children's books that introduce spatial sense are classified into one set; those books that simply identify shapes are another set; the third set extends geometric concepts and explores properties among geometric ideas. Other books that are related incidentally to geometry are organized into four sections—quilting, origami, distortions and optical illusions, and drawing.

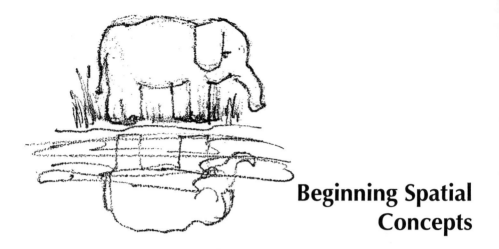

Beginning Spatial Concepts

Over and under, up and down, and in and out are some of the spatial concepts in these beginning concept books. Both Hoban's and McMillan's photographs capture these spatial ideas as well as build connections to our world. In *Becca Backward, Becca Frontward,* McMillan presents the concepts as pairs of opposites, whereas Hoban's photographs require the reader to identify and name the concepts in the everyday scenes. Readers will analyze Felix's charming illustrations to discern the opposites introduced in each "torn" frame. Beginning volume concepts are encountered as a reader tries to decide what is *inside* each of Carter's cleverly designed boxes; additionally the intriguing geometric shapes will captivate the reader. Books by Fanelli and Griffiths introduce another aspect of spatial sense: discerning how different objects are placed in a setting and in relationship to each other.

Allington, Richard L. *Opposites.* Illustrated by Eulala Conner. Milwaukee: Raintree Publishers, 1979.
See Measurement: Beginning Measurement

Carter, David A. *More Bugs in Boxes.* New York: Simon & Schuster Books for Young Readers, 1990.
ISBN 0-671-69577-0 $11.95
★★ Single concept, PS–2

Color and descriptive language are the focus of the sturdy pop-up book that is a sequel to Carter's counting book, *How Many Bugs in a Box?* Again, Carter's clever design and unusual boxes will intrigue the reader. Mathematics is not the theme of the book, but the reader will soon be captivated with the unique shape of the boxes and their contents. What could fit in the "yard-long YELLOW box" or the "RED rectangle box"? In addition to a cube and rectangular prisms of various shapes and sizes, other box shapes include an oblique prism, a cylinder, and two hexagonal prisms.

Duerrstein, Richard. *In-Out: A Disney Book of Opposites.* New York: Disney Press, 1992.

ISBN 1-56282-266-7 $5.95

☆ Single concept, PS

In-Out has the same format as *One Mickey Mouse: A Disney Book of Numbers.* The famous characters in this sturdy, square-shaped board book are illustrated in bright colors with minimal detail. Donald Duck, standing in an *open* doorway, welcomes the reader. A simple outline of a *closed* door dominates the opposite page. One page shows a view of a purple Pluto above a bold green horizontal line; the opposite page depicts Pluto under the line. This layout is used to introduce *over* and *under.* The book's only text consists of the words to convey open, closed; over, under; up, down; front, back; tall, short; and in, out.

Fanelli, Sara. *My Map Book.* New York: Harper Collins Publishers, 1995.

See Number—Extensions and Connections: Statistics and Probability

Felix, Monique. *The Opposites.* Mankato, Minn.: Creative Editions, 1993.

ISBN 1-56846-002-3 $5.95

★★ Single concept, PS–2

Clever design is consistent across Monique Felix's Mouse Books. The cover of the square book has an irregularly shaped cutout through which the reader can see a brown mouse and a white mouse huddled together. These two "opposites" are our guides through this no-text book. Each page shows an illustration of a rectangular piece of the page being torn on three sides and opened like a door so that we can see behind the "torn" piece. On a stark white background, the only art work is the mouse and what is being seen through the door. On opposite pages the white mouse opens a door while the brown mouse is behind a closed door, or the white mouse walks through the open door inside a room while the brown mouse walks through the door to outside the home. Concepts that are not named but are introduced include small, big; up, down; front, back; one, many; and dark, light. Some of the illustrations may represent more than one set of opposites. This deceptively simple-looking wordless book is more challenging than it seems.

Griffiths, Rose. *Railroads.* Illustrated by Peter Millard. Milwaukee: Gareth Stevens Publishing, 1995.

See Series and Other Resources: First Step Math.

Hill, Eric. *Spot Looks at Opposites.* New York: Putnam Publishing Group, 1986.

ISBN 0-399-21681-2 $3.75

☆ Single concept, PS–K

Full, empty; in, out; over, under; up, down; soft, hard; fast, slow; and open, shut are the opposites illustrated by the activities of the dog Spot and his animal friends. Brightly colored, simple illustrations and a minimal text clearly depict each concept. This irregularly shaped, sturdy cardboard book should appeal to the young child.

Hoban, Tana. *All about Where.* New York: Greenwillow Books, 1991.
ISBN 0-688-09697-2 $13.95
ISBN 0-688-09698-0   (library binding)   13.88
☆ Single concept, PS–3

Above, on, behind, under, out, against, across, between, in, through, beside, among, below, over, and around are the ideas explored in colorful photographs. The words are listed vertically along the outer borders of the first and last pages of this beginning concept book. Hoban's photographs are reproduced on narrower pages so that the list of concepts is always next to each new scene. Whether the photograph is of a child riding in a baby seat on a bicycle, children playing on a merry-go-round, a dog peeking his head out of a jacket pocket, or a ferry passing under a bridge, each new scene offers many options to be explored. The opportunities for communication about these concepts are many; children will be challenged to see how many of the relationships can be found in each picture.

———. *Over, Under, and Through and Other Spatial Concepts.* New York: Macmillan Publishing Co., 1973.
ISBN 0-02-744820-7 $11.95
ISBN 0-689-71111-5   (paper, Aladdin)   3.95
★★★ Single concept, PS–2

Tana Hoban's superb photographs explore the environment. Over, under, through, on, in, around, across, between, beside, below, against, and behind are some of the spatial concepts explored in Hoban's photographic essay. Children are spotlighted in everyday situations and positions as they lean against, run across, and climb through. The appealing black-and-white photographs focus so clearly on each spatial concept that children will have no difficulty comprehending.

———. *Push, Pull, Empty, Full: A Book of Opposites.* New York: Macmillan Publishing Co., 1972.
See Measurement: Beginning Measurement

Hutchins, Pat. *Changes, Changes.* New York: Macmillan Co., 1971.
See Geometry and Spatial Sense: Geometric Concepts and Properties

MacDonald, Suse, and **Bill Oakes.** *Puzzlers.* New York: Dial Books for Young Readers, 1989.
ISBN 0-8037-0689-8 $13.95
ISBN 0-8037-0690-1   (library binding)   13.89
☆ Single concept, PS–1

As in *Numblers,* the numerals from 1 through 10 are transformed and combined into attractive, imaginative pictures. On the left page of each two-page spread, three pictures illustrate the gradual change of each numeral according to some attribute, such as tall, backward, or upside-down. On the opposite page, various stylized numerals have been arranged into a picture of an animal. The reader must search the illustrations of the toucan, frog, or beaver to find examples of numerals that are tall, backward, or upside-down. A key to the pictures is in the back of the book, since some compositions are quite abstract. The young reader should be delighted with the imaginative shapes and challenged to find the numerals exemplifying the different attributes.

McMillan, Bruce. *Becca Backward, Becca Frontward.* New York: Lothrop, Lee & Shepard Books, 1986.
ISBN 0-688-06282-2 $16.00
ISBN 0-688-06283-0   (library binding)   15.93
★★ Single concept, PS–1

Cheery, colorful photographs of four-year-old Becca in her country environment illustrate this beginning concept book of opposites. The photographs of Becca filling a glass with milk and then drinking it clearly represent the picture captions of full and empty. Another pair of photographs show Becca playing with different-sized balls to indicate small and big. Other concept pairs include bottom and top, above and below, whole and half, and narrow and wide.

Rogers, Paul. *Surprise, Surprise!* Illustrated by Sian Tucker. New York: Scholastic, 1990.
ISBN 0-590-44473-5 $9.95
☆ Single concept, PS–1

Wrapped in colorful paper and ribbons, ten presents of various shapes and sizes are shown on an opening page with an invitation to the reader to guess what is inside. Hints are given in the rhyming text: "What's in the orange one? / It must be small. / An apple, perhaps? / Or a mug? / Or a ball?" These questions focus on the shape and relative size of each package. On each page, the reader can "unwrap" one or two presents to check his or her guess by lifting the movable flap. Careful observations are needed to keep track of how many objects have been unwrapped; three of the presents, a teddy bear, a scarf, and a sweater, are later pictured as a teddy bear wearing the scarf and sweater.

Time-Life for Children Staff. *Pterodactyl Tunnel: Amusement Park Math.* Alexandria, Va.: Time-Life for Children, 1993.

See Series and Other Resources: I Love Math

Shapes

Tana Hoban's *Shapes, Shapes, Shapes; Look Up Look Down;* and *Spirals, Curves, Fanshapes, & Lines* were three of the picture books that focus on geometry and its connections that Michele Staker decided to use with her fifth graders. Hoban's photographs of real-world examples of polygons and polyahedra in architecture and the world around us were used to introduce the study of two- and three-dimensional shape. Class discussions centered on the characteristics of the shapes.

With cameras donated by the Elementary Booster Club, the students took turns taking the cameras home to photograph polygons and polyhedrons in their neighborhoods. They wrote about their results in their journals; the discussions and their writing included analyzing and describing shape and using appropriate vocabulary. A discussion evolved on building design, which included energy efficiency, aesthetics, and multistory structures on a crowded block.

Four-year-old Sara also developed a fondness for *Spirals, Curves, Fanshapes, & Lines.* Her mother, Elana Joram, reports:

We read the book about five times, and each time, Sara became quite excited about identifying the pictures showing spirals. This was a new shape for her. In the past I have collected several things that are spiral shaped, and she seemed to pick up on this. She said that she knew where a spiral was and got my ammonite fossil, which is spiral shaped, from another room. Sara has shown a lot of interest in identifying spirals and figuring out what their distinguishing features are. We also talked about the differences between her drawings of hair, which tend to be circular, and a spiral.

Learning to recognize and name two- and three-dimensional shapes is the focus of this category. One-dimensional ideas are limited to a few examples of curved or broken lines. The general concept of shape with regard to form is explored by Seuss's *Shape of Me and Other Stuff* and Hoban's *Black on White, White on Black,* and *Who are They?* Some authors, such as Karlin and Smith, simply focus on one geometric two-dimensional shape. Others, such as Carle in *My Very First Book of Shapes,* explore many different shapes. Two- and three-dimensional shapes are often presented in the same book. Reiss and Smith are two authors who successfully relate two-dimensional shapes to the surfaces of three-dimensional shapes. Carle's *Draw Me a Star* can lead to an exploration of drawing star polygons; Ehlert's *Color Farm* and *Color Zoo* may motivate children to construct their own cutout book of shapes. Illustrations of many real-life examples of each shape are featured in most books. Connections to the world around us are effectively made in the photographic essays of Pluckrose and Hoban. MacKinnon uses photographs to focus explicitly on the *young* child's world. The types of experiences in these books can lead to a discussion of shapes in a child's world.

These beginning books often oversimplify concepts. They are not designed to explore the different properties of a shape or the interrelationships among shapes. Squares do not appear in the category of rectangles, distinctions are not made between a shape and its interior, and most figures are placed so their bases are parallel to the bottom edge of the page. These beginning experiences may cause children to overgeneralize. Confusion may also result from using illustrations of three-dimensional objects to represent two-dimensional shapes. Was it the author's intent to consider only the outline? Is the young child able to discern the difference? In spite of these limitations, these books can serve as an introduction to shape.

The following books are out of print but are considered exceptional:

Thomson, Ruth. *All about Shapes*

Yenawine, Philip. *Lines*

———. *Shapes*

Copies may be available in your library. See the first edition of *The Wonderful World of Mathematics* for annotations.

Allington, Richard L. *Shapes*. Illustrated by Lois Elhert. Milwaukee: Raintree Publishers, 1985.
ISBN 0-8114-8238-3   (paper) $3.95
★★ Single concept, K–2
Not only does this book introduce shapes—circle, square, triangle,

rectangle, oval, pentagon, hexagon, heptagon, and octagon—it also deals with the concept of size. The reader is asked to estimate which is the smallest pentagon illustrated, or the biggest hexagon, or to determine which shapes can fit into another shape. The book is illustrated in strong colors. Each shape is first shown in black, is named and defined, then is illustrated in full color as part or parts of large, black, silhouetted animals. The directions in the text lead the reader to make some perceptual judgments about the relative size of the shapes, which would be difficult to check. Three-dimensional shapes, such as spheres, cones, and cylinders, are briefly introduced; also, nonmathematical shapes, such as petals, hearts, stars, and crescents, are included. Finally, the reader is challenged to start a collection of shapes from items in the environment.

Arkadia. *Learning with Jemima Puddle-Duck: A Book of Shapes.* Avenue, N.J.: Derrydale Books, 1993.

ISBN 0-517-07699-3

☆ Single concept, PS

Based on the story and art of Beatrix Potter, this small board book introduces four shapes: oval, square, rectangle, and triangle. Each shape is simply introduced by an appropriate name printed underneath a black outline of the shape. The opposite page depicts the shape, highlighted with a black border, in the environment of Jemima Puddle-Duck. The pastel illustrations convey the charm of the character. The shapes are encountered as Jemima looks for a new house. The outlines depict a rectangular door, a square window, an oval egg, and the triangular outline of a roof. "When Jemima Puddle-Duck found her house, it had all the shapes she wanted"—a satisfying ending for the young reader.

Bishop, Roma. *Shapes.* New York: Little Simon, 1991.

ISBN 0-671-74830-0 $2.95

☆ Single concept, PS–K

Pop-up shapes in the forms of a square, a circle, a triangle, an oval, a rectangle, and a star appear as each cardboard page is turned in this tiny book. The only text is the appropriate name for the shapes that appear on each page. The objects shown are a mixture of two- and three-dimensional shapes. For a circle, for example, balloons, a sucker, and the sun are pictured; the rectangle is represented by a suitcase, a camera, and a house door. The objects are pictured straight on so that the three-dimensional qualities appear flat. The illustrations are in bright colors against a white background.

Bulloch, Ivan. *Games.* New York: Thomson Learning, 1994.

See Series and Other Resources: Action Math

Carle, Eric. *Draw Me a Star.* New York: Philomel Books, 1992.
ISBN 0-399-21877-7 $15.95

★★★ Single concept, PS–2

Drawing star polygons is the final feature in Carle's personal story from his childhood. In a handwritten letter at the book's end he relates how he learned to draw stars from his German grandmother and describes dreaming about shooting stars during vacations in his adult years. These two memories are incorporated into the simple text and collages in *Draw Me a Star.* Opening pages illustrate the step-by-step directions for drawing a five-pointed star; this drawing of "a good star" leads to a series of collage creations of the sun, a tree, a couple, a house, …, to finally a dark night, a full moon, and a new star, an eight-pointed one. Again each step is simply outlined. Carle's brilliant collages, simple text, and personal notes create a poignant story. *Draw Me a Star* can serve as a springboard into star polygons or geometric shapes. Note: Teachers will find the tasks described in the fourth-grade book from NCTM's Addenda Series a welcome resource for connecting star polygons to number.

————. *My Very First Book of Shapes.* New York: Crowell Junior Books, 1985.
ISBN 0-694-0013-2 $4.95

★★ Single concept, PS

Each cardboard page of this small book has been cut in half horizontally. Black silhouettes of a circle, a semicircle, a triangle, a diamond, a square, a rectangle, a crescent, and so on, are printed on the top half of each page. These shapes are to be matched with luxurious, painted illustrations of a kite, a moon, a tepee, an eye, a watermelon slice, a ladybug, a framed picture, and so on, on the bottom half. The half pages are randomly arranged so that the reader has to flip back and forth to match the shapes. The book contains no text or naming of shapes.

Disney's Pop-Up Book of Shapes. New York: Disney Press, 1991.
ISBN 1-56282-019-2 $6.95

★★ Single concept, no age given

Familiar Disney cartoon characters share equal billing with circles, triangles, squares, rectangles, and ovals in a circus atmosphere. Informal, but incomplete, definitions and a color silhouette introduce each shape. The pop-ups are most effective. As one page is turned, Pluto jumps through a hoop that is described as a circle. Another pop-up unfolds into a cube whose sides are defined as squares. Some examples are not as carefully illustrated; balloons, balls, tires, cylinders, and hoops are shown on the layout representing circles. In spite of potential confusion about two- and three-dimensional shapes, this book is recommended for its appeal and clever design; adult supervision can minimize conceptual misunderstandings.

Dodds, Dayle Ann. *The Shape of Things.* Illustrated by Julie Lacome. Cambridge, Mass.: Candlewick Press, 1994.
ISBN 1-56402-224-2 $13.95
☆ Single concept, PS–K

Shapes are everywhere! Not just on the end pages, not just in the borders, not just on the pages, either, but also all around us. The repetitive text invites children to chime in on repeated readings as the author sets the pattern: "A square is just a square, / Until you add a roof, / Two windows and a door, / Then it's much, much more." Each consecutive page introduces another shape: a circle, a triangle, a rectangle, an oval, a diamond. All are just shapes "Until you add...." Finally, a summary page places all the added-to shapes together in a double-page scene. Although the book challenges the reader to be more perceptive in observing the shapes around us, it does nothing to explain that the shapes are simply two-dimensional views. In some of the drawings, the oval becomes an egg and the square, a house. All the views are seen straight on, and a clear distinction is not made between three-dimensional shapes and their drawings. The paper-collage illustrations are charming, colorful, and childlike in style.

Dr. Seuss. *Shape of Me and Other Stuff.* New York: Random House, 1973.
ISBN 0-394-82687-6 $7.99
ISBN 0-394-92687-0   (library binding)   9.99
★★ Single concept, PS–1

Silhouetted shapes of a boy and a girl appear to bounce lightly from page to page as the children explore the shapes of things. Such concepts as "no shapes are ever quite alike" and "some objects have many shapes like gum and smoke" are cleverly illustrated. This is an intuitive exploration of shape rather than a study of explicit mathematical shapes that have unique forms because of their special properties.

Ehlert, Lois. *Color Farm.* New York: HarperCollins, 1990
ISBN 0-397-32441-3   (library binding) $12.89
ISBN 0-397-32440-5 $14.00
★★★ Single concept, PS–K

Color Zoo, a Caldecott Honor Book, featured Ehlert's clever design to introduce shapes. Ehlert uses the same approach to create another delightful book, *Color Farm.* Vibrant color, cutout graphics, and clever design effectively introduce shapes. Three pages, each with a differently shaped cutout, are layered over a fourth page.

The first graphic abstraction of a farm animal is labeled "rooster." When the reader turns the first page (with a square cutout), the opposite side of the page is labeled "square." The picture on the right-hand page is now labeled "duck." When this octagon cutout page is turned, the label is "octagon." The opening on the next page is hexagonal; all three shapes are named and shown on the fourth page. The flip side of each page introducing the die-cut shape's name is also one solid, bold color. Minimal graphics add limited detail to the abstract representations of farm animals. Other shapes that are introduced in a similar format are circle, heart, diamond, rectangle, oval, and triangle. Various colored shapes are used on the last pages to depict another representation of each of the eleven farm animals.

————. *Color Zoo*. Philadelphia: J. B. Lippincott, 1989.

ISBN 0-397-32259-3 $15.00
ISBN 0-397-32260-7   (library binding)   14.89
★★★ Single concept, PS–K

Bold color, cutouts, graphics, and clever design combine to introduce various shapes in a unique way. Three pages, each with a differently shaped cutout, are layered over a fourth page. The first graphic abstraction of a zoo animal is labeled "tiger." When the reader turns the first page (with a circular cutout), the opposite side of the page is labeled "circle." The picture on the right-hand page is now labeled "mouse." When this page with a square cutout is turned, the label is "square." The opening on the next page is triangular; all three shapes are named and shown on the fourth page. The flip side of each page introducing the die-cut shape's name is one solid, bold color. Minimal graphics add some detail to the zoo animals. Other shapes that are introduced are rectangle, oval, star, heart, hexagon, octagon, and diamond. The ten shapes, sixteen colors, and nine animals shown in the book are summarized on the last pages. Ehlert's clever design is very effective.

Grover, Max. *Circles and Squares Everywhere!* San Diego, Calif.: Browndeer Press, Harcourt Brace & Co., 1996.

ISBN 0-15-200091-7 $16.00
☆ Single concept, K–3

Grover's unique, colorful illustrations show circles and squares everywhere in an urban environment. Executed in a flat, cartoonlike style, the intensely colored illustrations become more complex with each page. An opening scene shows a tire shop along a highway; the opposite page has the simple text "Circles" printed inside a bright red circle. The next double page shows an overview of cars driving along a curvy city street through a neighborhood; the text reads "Tires and Cars" in a bold blue circle. The next pages become increasingly busy with the number of circular objects shown, the complexity of the city scenes, and the length of the simple text; the last scene is titled "Tires and Cars and Trucks and Roads" with an overhead view of a maze of twisting curves, ramps, and overhead lanes of a complex freeway. This same format is used to introduce squares. "Squares" is printed inside a red square; the opposite page shows a girl reading a book in a living room; squares are shown in the design of the chair's fabric, in the window panes, and in the picture frame. Increasingly complex overhead scenes show windows, houses, other buildings, and intersecting roads. Circles and squares are highlighted in the final third of the book. Grover's "flat" style is most appropriate for focusing on two-dimensional shapes in a three-dimensional world.

Hoban, Tana. *Circles, Triangles, and Squares.* New York: Macmillan Books for Young Readers, 1974.

ISBN 0-02-744830-4   (library binding) $13.95

★★★ Single concept, PS–2

Attractive black-and-white photographs of everyday scenes highlight the presence of circles, triangles, and squares. Some pictures feature only one shape but show it in different sizes, as in the frames and rectangular wire mesh of the rabbit cage or in the stack of pipes whose ends are circles of varied sizes. Other pictures include both circles (tires) and rectangles (frames) in the same scene. After studying these pictures, children should become more aware of their environment and the shapes within it. The photography is superb and stands alone without text.

————. *Black on White.* New York: Greenwillow, 1993.

ISBN 0-688-11918-2 $4.95

★★ Single concept, PS+

Silhouettes of an elephant, a pail, a maple leaf, and a ring of keys are easily identified and sharply defined against a white background in this small, sturdy board book. Rather than present specific geometric shapes, Hoban has captured the essence of the beginning study of shape.

————. *Blanco en Negro.* New York: Greenwillow, 1994.

ISBN 0-688-13653-2 $4.95

★★ Single concept, PS+

Tana Hoban does it again! Deceptively simple in format and attractive in design, this black on white sturdy board book features eight glossy pages of silhouetted items, some of which are familiar in a toddler's experiences. Free of clutter, the shapes jump clearly off each page, stimulating the viewer to identify them.

————. *Negro en Blanco.* New York: Greenwillow, 1994.

ISBN 0-688-13652-4 $4.95

★★ Single concept, PS+

See the review for Hoban's *Black on White*. Neither version contains text; consequently, the only difference between the books is whether the titles and information on the back cover are written in English or Spanish.

————. *Round and Round and Round*. New York: Greenwillow Books, 1983.

ISBN 0-688-01813-0 $14.95
ISBN 0-688-01814-9   (library binding)   14.88

★★★ Single concept, K–3

Hoban's colorful photographs clearly capture the concept of roundness. Two- and three-dimensional shapes (circles, spheres, and cylinders) are shown, such as wagon wheels, soap bubbles blown from a ring, and beads on a string. The scenes should be appealing to readers of any age. As in Hoban's other books, no text is given. The photographs invite discussion and increase the reader's appreciation of roundness in the environment.

————. *Shapes and Things*. New York: Macmillan Books for Young Readers, 1970.

ISBN 0-02-744065-5 $13.95

☆ Single concept, PS–2

Black-and-white images of common shapes in our world were made by a photogram process that does not use a camera. The distinct silhouette of each shape is clearly highlighted so that the differences among the shapes are accented. One layout of dinner flatware may cause the reader to speculate on the functions of the differently shaped forks, knives, and spoons. Only one layout depicts "pure" geometric shapes; the reader may wonder whether circles or spheres or cones or triangles are being photographed.

————. *Shapes, Shapes, Shapes*. New York: Greenwillow Books, 1986.

ISBN 0-688-05832-9 $16.00
ISBN 0-688-05833-7   (library binding)   15.93

★★★ Single concept, PS–3

Arcs, circles, hearts, hexagons, ovals, parallelograms, rectangles, squares, stars, trapezoids, and triangles are the shapes to explore in this wordless book. Colorful and beautiful photographs of the environment invite the reader to search for these shapes. The pictures explore shape in terms of both function and design. These examples can motivate readers to discuss and explore the world around them.

————. *Spirals, Curves, Fanshapes, & Lines*. New York: Greenwillow Books, 1992.

ISBN 0-688-11228-5 $14.00

★★ Single concept, PS–4

Tana Hoban takes us on a photographic exploration of the many shapes found in everyday objects in the world around us. More complex than her first books on this topic, like *Shapes, Shapes, Shapes,* this book highlights spirals, curves, and fanshapes and focuses more on shape as defined by line. Smoke curves upward as a young Asian child blows out the spiral candles on a birthday cake surrounded by curled paper streamers; a ram's strong spiral horns contrast with the gentle curve of his head. Perceptions are heightened and appreciation for the ordinary and beautiful is expanded. The photographs provide a delightful journey for readers of any age to value the form and function of geometric shapes.

————. *White on Black.* New York: Greenwillow Books, 1993.

ISBN 0-688-11919-0 $4.95

★★ Single concept, PS+

See the review for Hoban's *Blanco en Negro.* Neither version contains text; consequently, the only difference between the books is whether the titles and information on the back cover are written in English or Spanish.

————. *Who Are They?* New York: Greenwillow Books, 1994.

ISBN 0-688-12921-8 $4.95

★★ Single concept, PS+

Black on white, the silhouetted shapes of five familiar animals and their young challenge the young child to identify them. This sturdy board book also encourages beginning counting, from one lamb through five ducklings.

Karlin, Bernie. *Shapes: 3 Little Board Books.* New York: Little Simon Books, 1992.

ISBN 0-671-74625-1 $6.95

★★★ Single concept, PS–K

Circle, Square, and *Triangle* are the individual titles of the three sturdy board books that are distributed as a package. Each is uniquely shaped so that when the book is opened, the shape of the page is a circle, triangle, or square and six colorful examples dominate the entire double page. A stylized sun in bright colors of yellow, red, orange, and lavender dominates the front and back covers of the semicircular-shaped book, *Circle.* Another layout unfolds to show a circular dart board with the concentric rings sharply defined with colors of yellow, red, blue, black, and white. Spheres are shown on two pages, but because the outlines of the beach ball and the earth coincide with the shape of the circular page, the emphasis is on the two-dimensional outline rather than on the three-dimensional shape. The cover of *Square* depicts a top view of a gift-wrapped square box. Inside illustrations include childlike drawings on a square chalkboard, a checkerboard, a child peering through a four-paned window, and tall buildings with square windows shown on a square television screen. A final page depicts an alphabet block whose square faces are outlined in color. The cover of *Triangle* unfolds to depict a snow-capped triangular mountain with evergreens scattered along its side. Other full-page examples include a church steeple, a capital letter A, a stepladder, a Christmas tree, and the outline of a tepee.

MacDonald, Suse. *Sea Shapes.* San Diego, Calif.: Gulliver Books, 1994.
ISBN 0-15-200027-5 $13.95

☆ Single concept, PS–1

Similar in format to her earlier books *Numblers* and *Puzzlers*, this book depicts geometric shapes that are transformed into sea shapes. On the left page of each two-page spread, the shape is named and three or four pictures depict the gradual change of that shape. Each shape is modified by size and position as other colors and graphics are added. For example, curves and thick, wavy green lines transform a rectangular beige background containing a single orange five-pointed star into a scene depicting seaweed growing on sand across which three starfish move. On the opposite page a collage of the sea floor depicts three starfish. A lone light-blue circle on a turquoise rectangle becomes smaller and part of a larger purple shape. As the shape acquires more detail in the fourth frame, a whale emerges. On the opposite page is a sperm whale splashing its nose above the turquoise water as it captures a meal. Semicircles become jellyfish; hearts become butterflies; and fish and spirals become snails. Other shapes include squares, triangles, fans, ovals, diamonds, crescents, and hexagons. On the closing pages, "Sea Facts" names each sea creature and plant depicted in the collages and presents interesting facts about them and their habits.

MacKinnon, Debbie. *What Shape?* Illustrated by Anthea Sieveking. New York: Dial Books for Young Readers, 1992.
ISBN 0-8037-1244-8 $10.99

★★ Single concept, PS–K

Similar in format to MacKinnon's *How Many?*, this beginning shape book is illustrated with brightly colored photographs of charming children at play. Each shape is introduced with a question, such as "What shape is Claudia's tunnel?"; a close-up view of a child crawling through a bright red tunnel; and a response, such as "It's round." On the opposite page are "more round things," such as a teething ring, a bowl, a clock, a tomato, and toy balls. This page is framed with a brightly colored border splashed with small white shapes. All the children and objects were photographed with a white background, which removes unnecessary clutter. Two- and three-dimensional objects are shown in the examples, such as a wedge of cheese and a conical party hat for triangles. Many of the objects are photographed so that a two-dimensional quality is seen. The view of Oliver's sandbox was taken from above, which makes the sandbox appear square; the child's blocks predominantly show one face rather than the three-dimensional shape. Concepts introduced include round, square, triangle, rectangle, oval, heart, star, and diamond. The cast of children from various ethnic backgrounds is introduced on the end pages; each child is individually framed in the shape that she or he introduces in the book.

McMillan, Bruce. *Fire Engine Shapes.* New York: Lothrop, Lee & Shepherd Books, 1988.

ISBN 0-688-07842-7	$12.95
ISBN 0-688-07843-5 (library binding)	12.88

☆ Single concept, PS–2

Color photographs depict four-year-old Kaori's exploration of the details and shapes on a fire engine. Each pair of colorful close-ups clearly illustrates one dominant shape, such as the circular path of a handle for opening windows and the lug bolts on the wheels. Many opportunities for discussion of the familiar shapes exist. Squares, rectangles, diamonds, triangles, hexagons, ovals, and circles are defined on the last page beside their black silhouetted shapes. For easy reference in this photographic essay, the page number where each shape is featured is included. The definition of a diamond (rhombus) limits it to shapes that stand on a point, and the pictorial examples of a hexagon and a triangle are limited to regular polygons.

Morgan, Sally. *Circles and Spheres.* New York: Thomson Learning, 1994.

ISBN 1-56847-235-8

☆ Single concept, (1–3)

Circles and Spheres, like the other books in The World of Shapes series, has great potential but is uneven in quality. Color photographs clearly connect the shapes to the world around us. Diverse examples include scenes from nature and industry, neighborhoods and foreign lands, as well as products in retail stores and homes. Each shape is informally introduced through examples and descriptions; three-dimensional concepts are defined after two-dimensional shapes. The introduction to a circle includes a circular picture of the moon over a body of water; another circular picture shows a display of tomatoes. A third picture shows a boy holding a sphere. These are circular objects, but the caption "What is a circle?" is misleading. Parts of a circle are clearly illustrated as young children measure a tree to determine its circumference, and in another outdoor scene the semicircular markings on a basketball court are clearly captured. On the same page the reader is asked to measure the diameter and radius of a bright red circle, but the reader will first need to determine the circle's center. Descriptions are not mathematically correct. A sphere is inappropriately defined

as a solid circle; later a bubble is noted as being a hollow sphere; in
the glossary an oval is defined as "shaped like a stretched sphere."
Explorations of tiling a plane with circular tiles or filling a space
with spheres are excellent discovery tasks to learn about area and
volume as well as about properties of circles and spheres. Other
explorations include making a spinning top and finding circles on a
beautiful design. The circular ends of cylinders are depicted in
colorful photographs of spools of thread and tape and cans of soda
pop and food. The errors throughout this series could have been
readily avoided. Accurate definitions and explanations can be given
in informal language. This series' strength is in its real-world
examples, which can be used to appreciate the role that shape plays
in the world around us. A table of contents, a glossary, an index, and
a list of associated readings are included in each book in this series.

————. *Spirals.* New York: Thomson Learning, 1995.

ISBN 1-56847-278-1

★★ Single concept, (1–3)

Telephone cords, ringlets of hair, a corkscrew, the wire binding of
a notebook, and the threads of jar lids and wood screws are some of
the many examples of spirals that are part of our world. Spirals are
shown as staircases for lighthouses and fire escapes. Spirals are seen
in the tendrils of climbing plants, the uncurling of fern leaves, the
growth patterns of aloe, or the arrangement of a sunflower's seeds.
Examples from the animal world include the shell of a snail, the coil
of a snake, or the pattern of a spider's web. Connections to nature
also include the spiral of a tornado and the spiral of water as it flows
down a drain. Color photographs and descriptions of these items
build strong connections to spirals and their use. Directions for
making a spiral from sturdy paper are included. See the annotation
of Morgan's *Circles and Spheres* for an overview of the series, but
note that mathematical inconsistencies were not found in this book.

————. *Squares and Cubes.* New York: Thomson Learning, 1994.

ISBN 1-56847-234-X

☆ Single concept, (1–3)

Photographs of modern structures, examples in bright colors, and
descriptions introduce square and rectangular shapes. The
description for a rectangle incorrectly states that it is "a shape with
two long sides and two shorter sides." Although this describes the
examples given, it excludes the ideas that squares are also rectangles
or that the equal-length sides need to be opposite each other. Tiling
patterns based on a shape's properties are beautifully illustrated in
mosaics, stained-glass windows, and the design of bricks forming the
sidewalk by the Sydney Opera House. Other connections to the world
around us include shapes in games like Scrabble, dominoes, or table
tennis; in objects from our homes, such as cereal boxes, books, and

sugar cubes; or in the streets, such as traffic and advertising signs. A square traffic sign that is not resting on its base is referred to as "diamond-shaped"; it is unclear whether this statement implies that the sign is not also a square. The definition for a diamond (rhombus) included in the glossary does not clarify the confusion. A cube is well described as having six faces that are the same size and all squares, but inaccurate descriptions include "a cube is a solid shape" and "a brick is a solid rectangle with six sides." The terms *rectangular boxes, solid rectangles,* or simply *rectangles* are used rather than *rectangular prisms.* Activities for the reader include making a cube from a pattern and examining a colorful pattern of triangles to find the squares. Some of the challenges are difficult to meet because of the small size of the photographs. See the annotation of *Circles and Spheres* for an overview of The World of Shapes series.

————. *Triangles and Pyramids.* New York: Thomson Learning, 1995. ISBN 1-56847-277-3

☆ Single concept, (1–3)

Triangles are abundant in the color photographs of a stepladder, a traffic sign, the structure of a modern glass building, and a formal garden that introduce the concept of triangles through illustrations and descriptions. Triangular pieces become puzzles for forming a square from four congruent triangles and a hexagon from six congruent triangles. An easily avoidable error occurs when the hexagon is described as "a shape that has six equal sides." The puzzle pieces do form a regular hexagon, but the wording is careless. Two pages later, two more errors occur: "A solid triangle is called a pyramid. There are two types of pyramids, a square pyramid and a triangular pyramid." The triangular pyramid with four faces of the same size and shape is shown as a tetrahedron. Photographs depict the triangular shape in the musical instrument called a triangle, a tripod, the sails of boats and hang gliders, a shark's fins, and the wings of the Concorde. More traffic signs are shown, but one is called an upside-down triangle a label that does not indicate that a triangle is a triangle no matter what its position. The strength of the triangle is illustrated in the structures of bridges, pylons, and cranes; its beauty is noted in specific structures, such as the Eiffel Tower and the pyramids of Egypt. A colorful pattern for building a square pyramid is illustrated. See the annotation of *Circles and Spheres* for an overview of the series.

Pienkowski, Jan. *Shapes.* New York: Little Simon, 1987. ISBN 0-671-68135-4 $2.95

☆ Single concept, PS–K

Each new shape with its name is presented on one page of a two-page spread in this tiny cardboard book. The opposite page is a simple, boldly colored scene in which the shape is featured in several ways. The reader is introduced to circles, squares, triangles, rectangles, ovals, stars, crescents, spirals, waves, and zigzags. There is no attempt to distinguish among one-, two-, and three-dimensional shapes. An oval is pictured as a chicken egg. Spirals and zigzags, properties of lines, are depicted as curled fern fronds, snails, lightning bolts, and evergreen trees. Originally published in 1973, this Pienkowski book is reprinted as a nursery board book in an 8.5-cm-by-8.5-cm format.

Pluckrose, Henry. *Shape.* Illustrated by Chris Fairclough. Chicago: Children's Press, 1995.
See Series and Other Resources: Math Counts

Price, Marjorie. *1 2 3 What Do You See?: A Book of Numbers, Colors, and Shapes.* Stamford, Conn.: Longmeadow Press, 1995.
See Early Number Concepts: Counting—Bar Graphs

Puppy Round and Square. Illustrated by Norman Gorbaty. New York: Little Simon, 1991.
ISBN 0-671-74436-4 $3.95
Single concept, PS

Puppies scamper across the ten pages of this small, sturdy cardboard book. As the puppies pass trees and flowers, sailboats and ships, birds and kites, the text notes that all the objects around them have shape. Round and square are the two specific shapes introduced. For round, a ball is the example; for square, blocks are used. In the illustrations, the ball and blocks are shown as two-dimensional shapes. The simple text and cute illustrations are designed for preschoolers. An appealing aspect is the book's puppy shape.

Reiss, John J. *Shapes.* New York: Simon & Schuster Children's Books, 1982.
ISBN 0-02-776190-8 $13.95
ISBN 0-689-71121-2   (paper)   5.95
☆ Single concept, PS–1

Bold colors and bright graphics introduce the shapes of squares, triangles, circles, rectangles, and ovals; interesting designs dramatically show each shape in various sizes. Some images show the outline of the shape, but most representations include the interior. Real-life examples are used; for squares, the examples are crackers, a checkerboard, signal flags, and windows. Squares, triangles, and circles are also related to three-dimensional shapes. A very effective sequence shows a fox and a mole placing squares one at a time to form the six faces of a cube. Similarly, four triangles form the surfaces of a pyramid. The format for forming the sphere changes. The representation that relates circles and spheres is confusing because circles are stacked but it is not clear that the circles represent the cross section of the spheres rather than the surface. Examples of regular pentagons, hexagons, and octagons are included, as well as pages challenging the reader to find the various shapes in a scene formed from a collection of shapes.

Rogers, Paul. *The Shapes Game.* Illustrated by Sian Tucker. New York: Holt Books for Young Readers, 1990.

ISBN 0-8050-1280-X $12.95

☆ Single concept, PS–2

Bright colors dominate this introductory shape book. Different examples of a particular shape are introduced in verse and image. Similar shapes can be found in Matisse-like artwork on the facing page. Because the shapes are not identical and more than one of most examples are embedded in the collage, the reader should enjoy the challenge. Three-dimensional shapes are listed with two-dimensional names. For example, the circles named are a bubble, a round ball, a wagon wheel, the sun, and a drum. The drawings are flat; consequently spheres and cylinders look like circles. Careful attention needs to be taken to ensure that children realize that it is the outline of the drawing that is a circle. Although the graphics emphasize the outline to indicate two dimensions, the inclusion of both two- and three-dimensional items may confuse young readers. Other shapes introduced include triangles, squares, stars, ovals, crescents, rectangles, spirals, and diamonds.

Shaw, Charles G. *It Looked Like Spilt Milk.* New York: HarperCollins Publishers, 1947.

ISBN 0-06-025566-8 $13.95
ISBN 0-06-025565-X   (library binding)   13.89
ISBN 0-06-443159-2   (paper)   4.95

★★ Single concept, PS–2

Images of various familiar shapes appear to be constructed from torn paper. Each white shape is shown as a silhouette on a solid blue background. Each shape is described by repetitive text, such as "Sometimes it looked / like a Tree. / But it wasn't a Tree." The text is also printed in white on the blue background. The book's final statement explains the mystery: "It was just a Cloud in the Sky." Hoban's silhouette-shaped books are similar in format to this earlier classic.

Smith, Mavis. *Circles.* New York: Warner Juvenile Books Edition, 1991.

ISBN 1-55782-366-9 $3.95

☆ Single concept, PS–2

Each of the Flip-Flop books has the same design. The cover of each small, square cardboard book has a cutout of the shape being highlighted. The subsequent twelve pages also have cutouts that focus on the shape. In *Circles,* the cutouts highlight such objects as circular eyeglasses, egg yolks, Christmas tree ornaments, and oranges. The use of cutouts emphasizes that the shape of the *outline* is a circle. The oranges are spheres, but their outline on the page is a circle. Two examples appear contrived—a mouse's ear and a baby's mouth—but the circular roller-skate wheels and other examples are most appropriate. The reader is encouraged to find other circular shapes.

————. *Squares.* New York: Warner Juvenile Books Edition, 1991. ISBN 1-55782-364-2 $3.95

★★ Single concept, PS–2

See the annotation for Smith's *Circles.* Examples for squares include windows, crackers, pockets, a dog tag, and a pat of butter. Illustrations effectively show that a square is also one of the faces of a cube. This example is often inappropriately shown in beginning shape books.

————. *Triangles.* New York: Warner Juvenile Books Edition, 1991. ISBN 1-55782-365-0 $3.95

☆ Single concept, PS–2

See the annotation for Smith's *Circles.* Examples include a boat's sail, an arrow's tip, a cat's ear, and a tepee.

Time-Life for Children Staff. *The House That Math Built: House Math.* Alexandria, Va.: Time-Life for Children, 1993.
See Series and Other Resources: I Love Math

————. *Look Both Ways: City Math.* Alexandria, Va.: Time-Life for Children, 1992.
See Series and Other Resources: I Love Math

————. *The Mystery of the Sunken Treasure: Sea Math.* Alexandria, Va.: Time-Life for Children, 1993.
See Series and Other Resources: I Love Math

————. *See You Later Escalator: Mall Math.* Alexandria, Va.: Time-Life for Children, 1993.
See Series and Other Resources: I Love Math

Wegman, William. *Triangle, Square, Circle.* New York: Hyperion Books for Children, 1995.
ISBN 0-7868-0104-2 $6.95

☆ Single concept, PS–K

Parallel in format to his counting book *One, Two, Three,* this book by artist Wegman uses color photographs of weimaraners to illustrate geometric shapes. Each shape is introduced with the word name and a wooden block of the shape balanced on a dog's head. The opposite page shows the shape presented in a different context. For a square, a weimaraner is framed in a square opening of a box. Circles are shown as various-sized hoops around a weimaraner's neck, body, and legs; a triangle is shown as the sail on a boat in which a dog is sitting. Other shapes introduced include rectangles, semicircles, arches, and stars.

Geometric Concepts and Properties

Shelley Staker's fifth graders gave an overwhelmingly positive response to Burns's *The Greedy Triangle*. In giving a three-star rating to the book, the students included comments such as, "Every time you add a side, you add an angle." The study of prefixes for different polygons led to generating a class list of other words that use these same prefixes. And many students recognized the adage "Be happy with who or what you are."

A large number of excellent books dominate these further excursions into geometry and spatial sense. Topics are explored and extended to help the reader develop a more complete concept about properties of, and relationships among, geometric shapes and ideas. Symmetries are the focus of Walter's books. Tessellations, or covering a plane, are the centerpiece of Friedman's *A Cloak for the Dreamer*. Hindley's *The Wheeling and Whirling-Around Book* moves beautifully from two- to three-dimensional shapes and uses poetry, artwork, and real-life applications. Ross's *Circles: Fun Ideas for Getting A-Round in Math* is an excellent collection of worthwhile geometric tasks with fascinating connections to other cultures and history. Eratosthenes's work and times are recreated by Lasky in *The Librarian Who Measured the Earth*. Captivating activities in geometry and spatial sense abound in *Anno's Math Games* and in Burns's *The I Hate Mathematics! Book* and *Math for Smarty Pants*. Both the I Love Math and the First Step Math series provide explorations for the young reader; Smoothey's books offer an expansive investigation into these same topics for the older reader. Puzzle activities by Tompert and Walter furnish opportunities to investigate combining and changing shapes. Children and adults should enjoy many of the books described in this cate y.

The following books are out of print but are considered exceptional:

Adler, David A. *Three-D, Two-D, One-D*
Charosh, Mannis. *Ellipse*

Diggins, Julia E. *String, Straightedge, and Shadow: The Story of Geometry*
Froman, Robert. *Angles Are Easy as Pie*
———. *Rubber Bands, Baseballs, and Doughnuts: A Book about Topology*
Laithwaite, Eric. *Shape: The Purpose of Forms*
Orii, Eiji, and **Masako Orii.** *Simple Science Experiments with Circles*
Phillips, Jo. *Exploring Triangles: Paper-Folding Geometry*
———. *Right Angles: Paper-Folding Geometry*
Sitomer, Mindel, and **Harry Sitomer.** *Circles*
———. *What Is Symmetry?*
Testa, Fulvio. *If You Look around You*
Trivett, Daphne H. *Shadow Geometry*

Copies may be available in your library. See the first edition of *The Wonderful World of Mathematics* for annotations.

Adler, Irving. *Mathematics.* Illustrated by Ron Miller. New York: Doubleday, 1990.
See Number—Extensions and Connections: Number Theory, Patterns, and Relationships

Anno, Mitsumasa. *Anno's Math Games.* New York: Philomel Books, 1987.
See Number—Extensions and Connections: Number Theory, Patterns, and Relationships

———. *Anno's Math Games III.* New York: Philomel Books, 1991.
ISBN 0-399-22274-X $19.95
★★★ Multiconcept, 1–4 (2–5)

As pictures drawn on grids are stretched and shrunk from top to bottom or from side to side, Kriss and Kross explore what remains the same and what changes. This analysis gradually leads to more abstract sketches that are compared by using elementary topological ideas. A second chapter explores triangles through puzzle pieces, tessellations, and paper folding to form three-dimensional figures. In a third chapter, mazes are analyzed to solve the Königsberg bridges and other network problems. Mazes are reduced to simpler forms so that readers can learn how to draw and solve such problems. The relatively simple concepts of left and right gradually become more complex in the final chapter. Soon the reader is involved in deciding what is right or left for people in different positions and in determining how to give directions along a path. The afterword connects the concepts in the simple text and illustrations to more sophisticated mathematics and applications. The format is the same as in Anno's other game books, and like the others, this book is a delightful introduction to mathematical ideas.

Audry-Iljic, Françoise, and **Thierry Courtin.** *Discover Shapes.* Adapted by Judith Herbst. Hauppauge, N.Y.: Barron's Educational Series, 1995.

ISBN 0-8120-6499-2 $12.95

★★★ Single concept, PS (1–3)

Highly recommended for a number of ages is this unusual shape book, translated from a French edition, that explores one-, two-, and three-dimensional shapes. Illustrations are simply done with minimal detail in bold colors against stark backgrounds. The well-written text is in poetry form with irregular rhyming schemes. Bold print highlights the name of each new mathematical concept. Each concept is informally introduced through verse and illustrations and is followed by connections to the world around us. For example, the opening page states, "A **point** is very, very small, / It's hardly there at all." A lone small dot appears on the large, bold yellow page; the following page shows a child contemplating the stars against a dark purple sky. Many points drawn close together are used to introduce a line, which is later related to a column of little ants marching in a line. Smooth transitions and good examples continue with examples of straight, wavy, and broken lines. Curves and spirals lead to the introduction of a circle; an accompanying illustration depicts two children drawing a circle with a string anchored to a center stake. Other two-dimensional examples include disks, ovals, squares, rectangles, triangles, and diamonds. An illustration of a child rolling up a page of this book to form a cylinder cleverly marks the transition to three-dimensional shapes. Real-life cylinders include a flute, drinking straw, steam roller, rolling pin, and wallpaper rolls. Other shapes explored are spheres, cubes, pyramids, and cones. The foreword includes a note to the adult reader that describes why the study of shape is important and what young children know about shape. Children of different ethnic backgrounds appear throughout the book.

Bateson-Hill, Margaret. *Lao Lao of Dragon Mountain.* Illustrated by Francesca Pelizzoli, Chinese text by Manyee Wan, paper cuts by Sha-Liu Qu. London: De Agostini Children's Books, 1996. Distributed by Stewart, Tabori & Chang, New York.

ISBN 1-899883-64-9 $12.95

★★ Single concept, K–3

Patterns for cutting some of the symmetrical shapes shown throughout the book are included so the reader can explore the Chinese art of paper cutting. The final pages describe how to use a pattern to cut a butterfly with one line of symmetry. For cutting other symmetrical designs like a flower or a snowflake, the directions explain first how to fold a circle into twelfths. The paper-cut patterns were designed by Sha-Liu Qu. Lao Lao, the main character, is an old peasant woman who makes beautiful paper cuts for the children in the village. When the greedy emperor hears of her work, he has her imprisoned in a cold, barren tower until she makes a chestful of jewels. The beautifully told tale is printed in English and Chinese; an explanation of some of the Chinese characters is given at the end of the book. Pelizzoli's intricate, graceful illustrations are stunning and complement the story. *Lao Lao of Dragon Mountain* is a beautifully designed book that combines many elements to introduce the Chinese culture.

Bulloch, Ivan. *Patterns.* New York: Thomson Learning, 1994.
See Series and Other Resources: Action Math

————. *Shapes.* New York: Thomson Learning, 1994.
See Series and Other Resources: Action Math

Burns, Marilyn. *The Greedy Triangle.* Illustrated by Gordon Silveria. New York: Scholastic, 1994.

ISBN 0-590-48991-7 $15.95

★★★ Single concept, 1–4

Functions of different polygons and their connections to the world around us are the focus of this story about shape; it is also a clever story about self-acceptance. The triangle leads a busy and fulfilling life, building structures, such as roofs and bridges, or being the shape of a sailboat's sail or the musical instrument. The triangle's favorite occupation is to "slip into the place when people put their hands on their hips" and listen to the latest news, which is always shared with friends. But seeking more in life, the triangle goes to the Shapeshifter and asks for one more side and one more angle. After exploring new roles as a quadrilateral, such as a baseball diamond, a checkerboard, a computer screen, and a window frame, the shape again becomes dissatisfied and returns to the Shapeshifter. This story line continues as it explores the shape's different roles in life as a pentagon and then as a hexagon. As its sides and angles increase, its friends no longer recognize the shape and the shape is no longer comfortable with its explorations. It revisits the Shapeshifter and asks to return to its original shape. On the front and end pages, a row of "flowers" in flowerpots is ordered with each stem longer than the previous one. The flowers, each of which represents a different shape starting with a triangle, a quadrilateral, and a pentagon, are arranged in order of increasing sides and culminate with a dodecagon and a circle. Golds, fuschias, lavenders, blues, and greens are some of the colors that compose the whimsical cartoonlike illustrations.

————. *The I Hate Mathematics! Book.* Illustrated by Martha Hairston. Boston: Little, Brown & Co., 1975.
See Number—Extensions and Connections: Number Theory, Patterns, and Relationships

————. *Math for Smarty Pants.* Illustrated by Martha Weston. Boston: Little, Brown & Co., 1982.
See Number—Extensions and Connections: Number Theory, Patterns, and Relationships

————. *Spaghetti and Meatballs for All! A Mathematical Story.* Illustrated by Debbie Tilley. New York: Scholastic Press, 1997.
ISBN 0-590-94459-2 $15.95
Single concept, 2+ (2–6)

One of the Marilyn Burns Brainy Day Books, *Spaghetti and Meatballs for All!* is a lively exploration of perimeter and area relationships set in the context of seating for a family reunion. The story unfolds as Mr. and Mrs. Comfort decide to host a dinner for their family and friends, but the number of attendees, including the hosts, soon swells to thirty-two people. Mrs. Comfort rents eight card tables and thirty-two chairs and plans an arrangement of eight single tables. Bedlam occurs as soon as the first guests arrive. Over the protestations of Mrs. Comfort, the guests begin to rearrange the tables so they can sit together. Two tables pushed together now seat six guests; eight tables pushed together can accommodate only twelve guests. Eventually the problem is resolved, and the guests sit down for the sixteen loaves of garlic bread, eight pounds of spaghetti, eight quarts of spaghetti sauce, and ninety-six meatballs. The pace of the situation and the humorous antics of the guests and their pets are reflected in Tilley's detailed cartoonlike illustrations. At the book's end, three pages, "For Parents, Teachers and Other Adults," describe the mathematics involved through text and diagrams and suggest extensions for children.

Clouse, Nancy L. *Mapas rompecabezas de los Estados Unidos.* New York: Henry Holt & Co., 1994.
ISBN 0-8050-3598-2 $5.95
★★ Single concept, 1–3

See the annotation below for this Spanish version of *Puzzle Maps U.S.A.*

————. *Puzzle Maps U.S.A.* New York: Henry Holt & Co., 1990.
ISBN 0-8050-1143-9 $5.95
★★ Single concept, 1–3

Puzzle pieces from maps of the United States are the focus of this book on geography and shape. The puzzle pieces of the states are from maps of different sizes and colors. Pieces are rearranged to form designs. The reader is asked to identify the shapes by the name of the state and to find all the pieces that are the same shape no matter what the size, color, or position. The author has successfully connected geography with mathematical tasks.

Emberley, Ed. *Ed Emberley's Picture Pie: A Circle Drawing Book.* Boston: Little, Brown & Co., 1984.

ISBN 0-316-23425-7 $15.95
ISBN 0-316-23426-5   (paper)   7.95

★★ Single concept, (1–4)

Circles cut into halves, fourths, or eighths are the basis for the artwork in this unique book. Lots of various-colored circles are used to form elaborate designs; a number of these are symmetrical. As in Emberley's other works, directions for some designs are given, from simple to complex; some patterns show intriguing changes that result from simply moving the positions or colors of some of the circular parts. Designs vary; some include whole circles and others involve parts of circles, like 3 fourths or 7 eighths. Emberley's examples are a springboard for explorations of geometric designs or fractions that can be readily adapted to the reader's interests and talents. This book has proved popular with both students and teachers; children have created their own circle designs and discussed the number of circles and fractional parts that were used.

Friedman, Aileen. *A Cloak for the Dreamer.* Illustrated by Kim Howard. A Marilyn Burns Brainy Day Book. New York: Scholastic, 1994.

ISBN 0-590-48987-9 $14.95

★★★ Single concept, 1–4

Of three sons, only two prove themselves capable of joining their father in his work as tailor to the Archduke. The third, Misha, is a dreamer who wants to travel the world. To fulfill a commission, Ivan designs for the Archduke a patchwork cloak made from rectangles; Alex uses squares for one cloak and triangles for another; Misha uses circles. The first three cloaks are sold to the Archduke, but the fourth one is unacceptable and is set aside because it is full of open spaces. A plan occurs to the tailor; by cutting each circle into a hexagon, the tailor and his two older sons are able to redesign the cloak into a magnificent one that will protect Misha from the wind. The altered cloak is given to Misha as a going-away present as he sets off to see the world and fulfill his dreams. This beautiful story, so skillfully crafted, is illustrated with luminous watercolors that reflect the warmth and perception of an accepting father. Notes for parents, teachers, and other adults at the back of the book explain the mathematics in the story and provide suggestions for extending children's learning.

Griffiths, Rose. *Boxes.* Illustrated by Peter Millard. Milwaukee: Gareth Stevens Publishing, 1995.
See Series and Other Resources: First Step Math

————. *Circles.* Illustrated by Peter Millard. Milwaukee: Gareth Stevens Publishing, 1994.
See Series and Other Resources: First Step Math

————. *Printing.* Illustrated by Peter Millard. Milwaukee: Gareth Stevens Publishing, 1995.
See Series and Other Resources: First Step Math

————. *Railroads.* Illustrated by Peter Millard. Milwaukee: Gareth Stevens Publishing, 1995.
See Series and Other Resources: First Step Math

Hansen-Smith, Bradford. *The Hands-On Marvelous Ball Book.* Illustrated by Katie Sally Palmer. New York: W. H. Freeman & Co., 1995.
ISBN 0-7167-6628-0 $16.95
★★ Single concept, (3-8)

Sculptor-turned-author Hansen-Smith presents different aspects of geometry in this exploration of the many "faces" of an unusual ball as it is transformed into numerous three-dimensional shapes. As it spins, the sphere becomes four smaller spheres, all touching one another and all contained within the outline of the original sphere. Jimmy, the guide in this adventure, tries to pull the four spheres apart, but they remain connected as he pulls and stretches them into equivalent topological shapes. Suddenly the four spheres return to their original arrangement, and the centers of the spheres are connected with orange line segments to form a tetrahedron. The six points where the spheres touch are connected with green line segments to form an octahedron, inscribed within the tetrahedron. These shapes are then related to four congruent tetrahedrons, stacked like puzzle pieces to form the larger tetrahedron. The effective use of color and a series of illustrations help the reader follow the changes in shape. To create these marvelous shapes, nine-inch paper plates, masking tape, and bobby pins are needed. Step-by-step instructions are given for seven different shapes created primarily with tetrahedrons or octahedrons. Many of the shapes are formed by joining tetrahedrons at the edges, vertices, or faces. The story is told in rhyme and includes appropriate mathematical terminology. The illustrations are a combination of geometric diagrams and cartoonlike characters drawn as ink outlines splashed with bright color. This book can be appreciated at many different levels. *The Hands-On Marvelous Ball Book* is a result of the author's current work with mathematics and art teachers and an earlier study of Buckminister Fuller's work.

Hindley, Judy. *The Wheeling and Whirling-Around Book.* Illustrated by Margaret Chamberlain. Cambridge, Mass: Candlewick Press. 1994.
ISBN 1-56402-490-3 $14.95
★★★ Single concept, PS–3 (2–6)

"Things that spin and whirl— / things that wheel and reel and roll / and curve and coil and curl"—Hindley's poetic text complements Chamberlain's whimsical illustrations to create an appealing approach to the study of circular objects. Carefully incorporated through art, interesting asides, and the main prose are facts about circles, cylinders, spheres, and spirals. This information includes how we use them, their occurrences in nature, and their mathematical properties. "You can race them or chase them or bowl them away"; these written descriptions are matched by illustrations of jesters, ladies-in-waiting, kings, and squires rolling such circles as coins, rings, and crowns along the rim of the earth. Included in the lively illustrations are short asides or activities, such as declaring a pizza is disk shaped, showing how to fold a paper circle to find its center, defining an axis, and exploring a circle's definition through a construction with a compass. A clever shift from circles to spheres is accomplished by considering the shape of a twirling circle; visual descriptions of spheres, circles, and arcs are depicted by a pirouetting ballerina, a spinning coin, a rolling dog, and a tumbling acrobat. Similar transitions consider a circle as a slice of a cylinder or a spiral as the "curling, coiling peel" from a cylinder's surface. Connections to the function of shape include the example of a cylinder, which can be a roller for making paper, the rollers on a conveyor belt, cans, pipes, cups, drums, and tube socks. Connections are also made to the shape of the universe, orbits of planets, and atoms. This delightful book, the result of creative and thoughtful design, should prove popular with readers of many ages.

Hutchins, Pat. *Changes, Changes.* New York: Macmillan Publishing Co., 1971.
ISBN 0-02-745870-9 $14.95
★★ Single concept, PS–1

Two wooden dolls arrange and rearrange wooden building blocks to make various objects. Each transformation forms a totally new shape from a set of blocks that includes cylinders, rectangular and triangular prisms, arches and sectors. The constructions represent a home, a fire engine, a boat, a truck, and a train as the characters reposition the blocks thoughtfully, sometimes frantically, but always at a fast pace. This introduction to building three-dimensional objects should inspire readers to create structures of their own. The illustrations are pen-and-ink drawings with colorful overlays of red, orange, blue, yellow, and green.

Lasky, Kathryn. *The Librarian Who Measured the Earth.* Illustrated by Kevin Hawkes. Boston: Little, Brown & Co., 1994.

ISBN 0-316-51526-4 $16.95

★★★ Single concept, 1–5 (2–6)

Eratosthenes was full of wonder and questions about the world around him. This picture-book biography celebrates his curiosity and his questioning mind as it describes how 2000 years ago, he discovered a mathematical way to measure the circumference of the earth. His calculations are within 200 miles of today's calculations, which are performed with highly sophisticated technology. His strategy included using 360 degrees in a circle, measuring angles, using ratios to calculate lengths with shadows, and measuring lengths with nonstandard means. Stubborn camels and bematists— surveyors who were trained to walk with equal-length paces—are included in this humorous description of his trials to measure the distance between two locations. In notes at both the beginning and the end of the book, Lasky describes how she became intrigued with Eratosthenes' work and sets a context for his discoveries with respect to both the culture and the known mathematics at that time. The text and illustrations are intriguing as they capture the detaile of ancient Greek life and describe the constant questioning and curiosity of this inspiring man. Lasky provides an excellent opportunity to develop a sense of history and to see mathematics as a human endeavor.

Pluckrose, Henry. *Pattern.* Illustrated by Chris Fairclough. Chicago: Children's Press, 1995.

See Series and Other Resources: Math Counts

Riggs, Sandy. *Circles.* Illustrated by Richard Maccabe. New York: Benchmark Books, 1997.

See Series and Other Resources: Discovering Shapes

———. *Triangles.* Illustrated by Richard Maccabe. New York: Benchmark Books, 1997.

See Series and Other Resources: Discovering Shapes

Ross, Catherine Sheldrick. *Circles: Fun Ideas for Getting A-Round in Math.* Illustrated by Bill Slavin. Reading, Mass.: Addison Wesley, 1992.

ISBN 0-201-62268-8   (paper) $9.95

★★★ Single concept, 3–9

Intriguing tasks, demonstrations, applications, explanations, and humor maintain the interest of the reader throughout this fascinating and in-depth exploration of the circle and other round shapes. Prepare to be riveted by the text and its examples as you increase your awareness of the properties of a circle, its connections to our world, and its historical significance. Some of the connections related in the chapter "Living in Circles" include examples of line and radial symmetry in nature and the circle signs designed by cave dwellers, meteorologists, and botanists; building with circles, including the circular huts of the Zulus, the defense plan of Arctic musk oxen, and the designs of Jerusalem, the Viking camp at Trellebrog, and many cities in fifteenth- and sixteenth-century Europe; and circle art in the Chinese yin and yang symbol and the Catherine wheel. Tasks in this section for the reader to pursue include printmaking with vegetables, cutting a super circle from a piece of paper, and using compass and straightedge to design circle petals and to create hexagons, triangles, and squares to explore tiling. In addition to the properties of circles, those of ellipses, spheres, disks, cylinders, cones, and spirals are considered in the nine chapters. This stimulating text is accompanied by line drawings and fresh watercolor characters as lively as the text. Drawings clarify instructions for each project. A glossary and an index are included. Ross's *Circles* is an exemplary book that is rich in connections, entertainment, and mathematics—a must for intermediate and junior high school teachers and readers!

Sharman, Lydia. *The Amazing Book of Shapes*. New York: Dorling Kindersley, 1994.

ISBN 1-56458-514-X $14.95

★★★ Single concept, K–3 (K–4)

Geometric patterns and shapes are the focus of another quality Dorling Kindersley publication. Excellent connections to the real world are captured in color photographs throughout the book. Each well-designed two-page spread focuses on one concept. For every concept, an accompanying project is clearly described through illustrations and step-by-step directions. The wide variety of tasks range from simple to complex. Centers of shapes are explored by constructing and folding a series of squares that are arranged to form a colorful design. Centers are also considered in designs formed by drawing overlapping circles with a compass; readers are encouraged to create their own designs with color and other constructions. In several examples, line symmetry can be explored with the mirror card attached to the ribbon bookmark. Rotational symmetry is well defined and illustrated through explanations and examples, including designs from art pieces. Another project involves the designs created by star polygons and star shapes formed from rotating polygons around a point. Other geometric concepts are explored through designing both linear and two-dimensional patterns, tessellating shapes, solving tangrams and other puzzle problems, constructing fractals, and building nets for three-dimensional figures. A sturdy foldout section includes carefully designed stencils of shapes that are essential for many of the projects. An introductory section includes notes to parents or other adults about the features of the book and indicates needed materials. A glossary titled "Words about Shapes" includes definitions of such terms as *rotational* and *mirror symmetry* and of terms for three-dimensional shapes from simple cubes to a complex-sounding rhombicuboctahedron. This exceptional book offers something new for everyone.

Simon, Seymour. *Mirror Magic.* Illustrated by Anni Matsick. Honesdale, Pa.: Bell Books, 1991.

ISBN 1-878093-07-X $ 9.95

★★ Single concept, 2–5

Simon is an award-winning author of many acclaimed science books for children. His interesting explanation of how a mirror works and his guided experiments and activities for children combine to make a stimulating text. Directions, simply stated and understandable, are given for making a kaleidoscope, a periscope, and a corner-cube mirror like those used in reflectors. Other simple suggestions will motivate children to experiment with angles and mirrors on their own. For example, as angles are changed between two or three hinged mirrors, children will be delighted and surprised at the changing images. Matsick's illustrations complement the text and carefully capture the reversals of mirror images.

Smoothey, Marion. *Angles.* Illustrated by Ted Evans. New York: Marshall Cavendish, 1993.

See Series and Other Resources: Let's Investigate

———. *Area and Volume.* Illustrated by Ted Evans. New York: Marshall Cavendish, 1992.

See Series and Other Resources: Let's Investigate

———. *Circles.* Illustrated by Ted Evans. New York: Marshall Cavendish, 1993.

See Series and Other Resources: Let's Investigate

———. *Maps and Scale Drawings.* Illustrated by Ann Baum. New York: Marshall Cavendish, 1995.

See Series and Other Resources: Let's Investigate

———. *Quadrilaterals.* Illustrated by Ted Evans. New York: Marshall Cavendish, 1992.

See Series and Other Resources: Let's Investigate

———. *Ratio and Proportion.* Illustrated by Ann Baum. New York: Marshall Cavendish, 1995.

See Series and Other Resources: Let's Investigate

———. *Shapes*. Illustrated by Ted Evans. New York: Marshall Cavendish, 1993.
See Series and Other Resources: Let's Investigate

———. *Shape Patterns*. Illustrated by Ted Evans. New York: Marshall Cavendish, 1993.
See Series and Other Resources: Let's Investigate

———. *Solids*. Illustrated by Ted Evans. New York: Marshall Cavendish, 1993.
See Series and Other Resources: Let's Investigate

———. *Triangles*. Illustrated by Ted Evans. New York: Marshall Cavendish, 1993.
See Series and Other Resources: Let's Investigate

Stienecker, David L. *Patterns*. Illustrated by Richard Maccabe. New York: Benchmark Books, 1997.
See Series and Other Resources: Discovering Shapes

———. *Polygons*. Illustrated by Richard Maccabe. New York: Benchmark Books, 1997.
See Series and Other Resources: Discovering Shapes

———. *Rectangles*. Illustrated by Richard Maccabe. New York: Benchmark Books, 1997.
See Series and Other Resources: Discovering Shapes

———. *Three-Dimensional Shapes*. Illustrated by Richard Maccabe. New York: Benchmark Books, 1997.
See Series and Other Resources: Discovering Shapes

Talbot, John. *It's for You: An Amazing Picture-Puzzle Book*. New York: Dutton Children's Books, 1995.
See Number—Extensions and Connections: Reasoning—Classification, Logic, and Strategy Games

Time-Life for Children Staff. *The Case of the Missing Zebra Stripes: Zoo Math*. Alexandria, Va.: Time-Life for Children, 1992.
See Series and Other Resources: I Love Math

———. *How Do Octopi Eat Pizza Pie: Pizza Math*. Alexandria, Va.: Time-Life for Children, 1993.
See Series and Other Resources: I Love Math

———. *Look Both Ways: City Math*. Alexandria, Va.: Time-Life for Children, 1992.
See Series and Other Resources: I Love Math

————. *Right in Your Own Backyard: Nature Math.* Alexandria, Va.: Time-Life for Children, 1993.

See Series and Other Resources: I Love Math

Tompert, Ann. *Grandfather Tang's Story.* Illustrated by Robert Andrew Parker. New York: Crown Publishers, 1990.

ISBN 0-517-57487-X $16.00
ISBN 0-517-57272-9   (library binding)   16.99

★★ Single concept, PS–2

Watercolor paintings illustrate this delightful story about fox fairies that Grandfather Tang shares with Little Soo. The essence of the adventurous tale of the two foxes, Chou and Wu Ling, is their challenge to outdo each other by changing themselves into different animals. Both foxes are pictured in the delicate illustrations, and each is represented by the seven tangram pieces. As the tale proceeds, they change into such forms as a rabbit, a dog, a squirrel, a hawk, and a crocodile. As each transformation occurs, the tangram pieces are rearranged to depict the new form. Readers can trace a copy of the tangram pieces so that they can explore the different shapes by using their own pieces.

Van Fleet, Matthew. *Match It: A Fold-the-Flap Book.* New York: Dial Books for Young Readers, 1993.

ISBN 0-8037-1379-7 $11.95

☆ Single concept, PS–3

Thirty puzzles involving spatial sense are presented in this sturdy little book. Six colorful figures are shown on the left-hand page of each of five double pages. Three full-sized pages are layered over the opposite page. Each fold-out page contains part of each figure and cutouts. The reader must decide in which order to refold the pages to form a specific figure. To match another figure on the page, the reader must reorder the pages; each solution is unique. Puzzles include pictures of toys, fruit, numerals, animals, and geometric shapes.

Vorderman, Carol. *How Math Works.* Pleasantville, N.Y.: Reader's Digest Association, 1996.

See Series and Other Resources: Resources

Walter, Marion. *Look at Annette.* New York: M. Evans & Co., 1977.

ISBN 0-87131-071-6 $5.95

★★ Single concept, PS–3

Children will delight in this adventure in discovery learning. A pocket on the inside cover of the book contains an unbreakable mirror. The reader is invited to meet the challenges of the text by experimenting with the placement of the mirror. No further explanation is given, leaving the responsibility of discussing the concept of symmetry to the teacher or other adult readers after the child discovers it through the explorations of a little girl's fun with her toys.

———. *Make a Bigger Puddle, Make a Smaller Worm.* New York: M. Evans & Co., 1970.
ISBN 0-87131-073-2 $5.95
★★ Single concept, PS–3

A mirror attached to the inner cover allows the reader to explore the questions on each page of this book. Interesting problem-solving situations using mirror reflection abound; namely, making a tower of blocks higher or wider; making a puddle bigger or smaller; and changing a set of three dots into six, five, or two dots. A simple text complements colorful illustrations in this creative approach to exploring symmetry.

———. *The Mirror Puzzle Book.* Jersey City, N.J.: Parkwest Publications, 1985.
ISBN 0-906212-39-1   (paper) $6.95
★★ Single concept, 2+

Mirror images of the book's title, the title page, and the directions for using the book cleverly introduce this colorful book on symmetry. As in her other books, Walter provides a mirror attached to the inner cover that the reader can use to create the desired designs. Each double page has twelve puzzle designs that the reader is to reproduce by using one basic design and the mirror. Depending on the visual-perception skills of the reader, several attempts at mirror placement may be needed before the correct image is obtained. The construction of the designs for the twelve puzzles varies in difficulty. Some of the 144 puzzle designs in the book are impossible; a scrambled chart at the end of the book lists the solvable puzzles. The reader is encouraged to predict which designs are solvable before using the mirror.

White, Laurence B., and **Ray Broekel.** *Math-a-Magic: Number Tricks for Magicians.* Illustrated by Meyer Seltzer. Morton Grove, Ill.: Albert Whitman & Co., 1990.
See Number—Extensions and Connections: Reasoning—Classification, Logic, and Strategy Games

Incidental Geometry— Origami

Explicitly following step-by-step directions and illustrations can transform a sheet of paper into a cube or a crane. Not only will readers find these paper-folding activities motivating, but as they ponder the diagrams and fold their papers, they will strengthen their spatial sense.

Anno, Mitsumasa. *Anno's Math Games III.* New York: Philomel Books, 1991.
See Geometry and Spatial Sense: Geometric Concepts and Properties

Araki, Chiyo. *Origami in the Classroom, Book I.* Rutland, Vt.: Charles E. Tuttle Co., 1965.
ISBN 0-8048-0452-4 $14.95
★★ Single concept, 2+
    Designed to teach origami to children in the United States, this book is well planned for use by teachers. The origami themes in this volume correspond to the fall school semester, with activities for autumn through Christmas. Each activity lists appropriate grade levels, materials needed, and estimated time for the activity; the estimates of time appear excessive. The labeled diagrams are illustrated in two colors with dotted lines and arrows to indicate folds and directions of folds. The step-by-step diagrams clearly complement the written text, although younger children may need adult assistance in following the written directions. The cross-cultural focus of these activities is excellent, although the idea of an origami Santa Claus is a little startling.

———. *Origami in the Classroom, Book II.* Rutland, Vt.: Charles E. Tuttle Co., 1968.
ISBN 0-8048-0453-2 $14.95
★★ Single concept, 2+

Activities for winter through summer are the theme of this second volume of origami by Araki. A colorful full-page illustration introduces each new set of directions for projects such as George Washington riding a horse and a box for valentines. See the annotation for *Origami in the Classroom, Book I* for a more complete review.

Murray, William D., and **Francis J. Rigney.** *Paper Folding for Beginners.* New York: Dover Publications, 1960.

ISBN 0-486-20713-7   (paper) $2.50

☆ Single concept, 1+

Photographs of completed projects and drawings of step-by-step, labeled diagrams illustrate more than forty projects involving paper folding. This is not a traditional origami book because some projects require more than one piece of paper and others involve cutting or tearing the paper. For some projects, the narrative refers the reader back to previous projects and figures. Although readers of many ages will be fascinated with the projects in this book, the directions are not as detailed as in other books. For young readers, adult direction will be needed.

Sakade, Florence. *Origami, Japanese Paper Folding.* Vol. 1. Rutland, Vt.: Charles E. Tuttle Co., 1957.

ISBN 0-8048-0454-0   (paper) $3.75

☆ Single concept, K–12

One of a series of three, this softcover book furnishes background on origami, a favorite pastime of Japanese children. Step-by-step illustrations and written directions detail the art of Japanese paper folding for fifteen different objects. All the objects are folded from square sheets of paper; each corner and fold is carefully labeled and described in appropriate mathematical language. The designs, which vary from simple folding of fans and sailboats to more complex folding of swans and cranes, reflect a good variety of traditional origami projects. The paper-folding designs are generally easy to do, and with assistance from a teacher or a parent, they would provide an appropriate level of challenge for a child who is new to the craft. The placement of the printed directions apart from the diagrams could be confusing. In the second and third books of the series, this problem has been overcome; a written direction is beside each step of the illustrated process.

————. *Origami, Japanese Paper Folding.* Vol. 2. Rutland, Vt.: Charles E. Tuttle Co., 1958.

ISBN 0-8048-0455-9   (paper) $3.75

★★ Single concept, K–12

See *Origami, Japanese Paper Folding,* Vol. 1, for a complete annotation of the three-book series.

———. *Origami, Japanese Paper Folding.* Vol. 3. Rutland, Vt.: Charles E. Tuttle Co., 1959.

ISBN 0-8048-0456-7   (paper) $5.95

★★ Single concept, K–12

See *Origami, Japanese Paper Folding,* Vol. 1, for a complete annotation of this series.

Sarasas, Claude. *ABC's of Origami.* Rutland, Vt.: Charles E. Tuttle Co., 1964.

ISBN 0-8048-0000-6 $12.95

★★ Single concept, 3–8

From albatross to zebra, origami patterns for twenty-six objects are illustrated with sequenced diagrams. For some of the more complex constructions, the diagrams are difficult for a novice to follow. There is minimal use of arrows to indicate the directions of the folds and only an occasional use of notation such as "front," and vertices are not labeled with letters. A full-color illustration of the completed paper folding and the object's name in English, French, and Japanese introduces each set of directions.

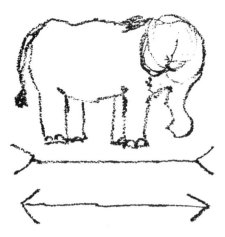

Incidental Geometry— Distortions and Optical Illusions

Are those lines really the same length? Is the open face of the cube on top or in the front? What is your perception? Is it the same from one moment to the next? Numerous familiar and unfamiliar illusions abound in these books where lines or color is effectively used to change the reader's perception. Readers are invited to create their own illusions by the Baums in *Opt: An Illusionary Tale* and by Supraner in *Stop and Look! Illusions*. Some books, such as White and Broekel's *Optical Illusions,* offer scientific explanations for the illusions.

Connections to art and geometry are found in books like Bolton's *Hidden Pictures* and Rand's edition of Carl Sandburg's *Arithmetic;* notes to the reader include discussions on how to use grids to draw such distortions. Spatial relationships can be analyzed in Jonas's work as a reader considers each illustration from two viewpoints. Hoban's *Shadows and Reflections* cause the reader to consider similarities in a shape and its reflection.

The following books are out-of-print but are considered exceptional:

Anno, Mitsumasa. *Topsy-Turvies: Pictures to Stretch the Imagination*
———. *Upside-Downers*

Copies may be available in your library. See the first edition of *The Wonderful World of Mathematics* for annotations.

Baum, Arline, and **Joseph Baum.** *Opt: An Illusionary Tale.* Markham, Ont.: Viking Kestrel, 1987.

ISBN 0-670-80870-9	(paper)	$11.95
ISBN 0-14-050573-3	(paper, Puffin Books)	4.99

★★★ Single concept, PS–3

Opt, the kingdom of optical illusions, is the setting for familiar illusions in colorful scenes of medieval castles and royalty. The story line queries the reader on which line is longer, whether the lines are parallel, which circle is larger, and how many cubes are in the stack. In addition to geometrical-contrast illusions involving angles, lines, and shapes, illusions involving color, impossible objects, and hidden figures add interest to the illustrations. The closing pages explain the rationale behind each illusion and suggest how readers can create their own illusions.

Bolton, Linda. *Hidden Pictures.* New York: Dial Books, 1993.

ISBN 0-8037-1378-9 $14.99

☆ Single concept, 4+

Art is the main theme of *Hidden Pictures,* but Bolton's unique selections contain excellent examples of geometric concepts. Some selections, such as Dali's *The Face of Mae West* or Picasso's *Three Musicians,* may simply intrigue or amuse. Other selections are more closely related to mathematics: tessellations can be seen in Escher's work; mirror reflection is involved in reading Leonardo Da Vinci's handwriting; several artworks are distorted or anamorphic paintings; some images can be clearly viewed only on the surface of a cylinder formed from the Mylar sheet enclosed with the book. Additional pages at the book's end include directions for making puzzling pictures by using reflections and distorted grids or by modifying angles and lines to create new images. Books like *Hidden Pictures* help readers discover the connections between mathematics and other topics. See the review of Ted Rand's edition of Carl Sandburg's *Arithmetic* for another example of a book that connects art and geometry.

Brandreth, Gyles. *The Great Optical Illusions.* Illustrated by Rowan B. Murphy and Albert Murphy. New York: Sterling Publishing Co., 1985.

ISBN 0-8069-6258-5   (paper) $4.95

☆ Single concept, 2+

Straight lines appear curved; congruent shapes appear different. Pictures change content before the reader's eyes; impossible constructions appear possible. This is an intriguing collection of classical illusions illustrated in crisp black and white.

Carini, Edward. *Take Another Look.* Englewood Cliffs, N.J.: Prentice-Hall, 1970.

ISBN 0-685-03910-2   (paper) $1.50

★★ Single concept, PS–3

Bold graphics in bright yellows, greens, and blues dramatically illustrate familiar optical illusions. Straight lines appear crooked; congruent lines and objects appear unequal. Contrast between colors causes similar colors to appear darker and other colors to "dance." A simple text invites readers to explore each illusion and check their perceptions.

Hoban, Tana. *Shadows and Reflections.* New York: Greenwillow Books, 1990.

ISBN 0-688-07089-2	$12.95
ISBN 0-688-07090-6 (library binding)	12.88

★★ Single concept, PS+

Distortions may appear when shapes are reflected in mirrors, windows, puddles, or ponds, but certain attributes are unchanged. Similarly, elongated or abbreviated images are found in shadows. Hoban captures these images in her stunning color photographs. This collection presents an excellent opportunity for the reader to analyze shape by comparing and contrasting objects with their shadows and reflections.

Jonas, Ann. *Reflections.* New York: Greenwillow Books, 1987.

ISBN 0-688-06140-0	$15.00
ISBN 0-688-06141-9 (library binding)	14.93

☆ Single concept, K–4

Jonas has carefully crafted her illustrations so that the book can be read front to back, turned over, and reread back to front. In *Reflections,* the illustrations are in color, and the text traces the activities of a child's day. The illustrations are a study of changing images and perspective; readers will want to restudy the pages to analyze each illusion. The sailboats in the bay under dark thunderclouds become kites framed against a pale blue sky and flown by family members standing on the sandy beach. The vacated boatyard by the blue lake becomes a campground under a blue sky.

————. *Round Trip.* New York: Greenwillow Books, 1983.

ISBN 0-688-01772-X	$16.00
ISBN 0-688-01781-9 (library binding)	15.93

☆ Single concept, K–4

Black-and-white illustrations and a simple text trace a family's day trip to the city and their return home that evening. Each silhouette is carefully crafted so that when the reader finishes the last page, he or she can turn the book over and reread it. This format is similar to that of Jonas's *Reflections.* The images are a study of lines and perspective; readers will want to reexamine the pages to analyze each illusion. The barns, silos, and curved furrows of a farmer's field become a factory complex with smoky fumes. The straight highways that lead into the skyscraper skyline against a white background become searchlights piercing the night. White print on a black background conveys the first half of the journey; the second half is recorded with black print on a white background.

Paraquin, Charles H. *World's Best Optical Illusions*. New York: Sterling Publishing Co., 1987.

ISBN 0-8069-6644-0   (paper) $4.95

☆ Single concept, 4–12

One hundred seven illusions are simply and effectively illustrated in black-and-white drawings. A question for the reader is posed about each diagram. The illusions are clustered with a brief narrative encouraging the reader to consider the different aspects of that section. In one section, the reader is challenged to find embedded figures in complex drawings. Spatial perception is also exercised in illusions involving comparisons and looking at figures from more than one point of view.

Sandburg, Carl. *Arithmetic*. Illustrated by Ted Rand. San Diego: Harcourt Brace Jovanovich Publishers, 1993.

ISBN 0-15-203865-5 $17.00

★★ Single concept, 3+

Carl Sandburg's familiar lines are dramatically interpreted through Rand's anamorphic images. "Arithmetic is where the numbers fly like pigeons / in and out of your head." The accompanying illustration to this opening line is an elongated image of numerals and pigeons that appear to fly across the page from a man's head. From a red arrow on the side of a page, a few lines drawn in perspective radiate outward. When the reader closes one eye and holds the book to view the page from the vantage point indicated by the red arrow, the images on the page no longer appear distorted. Anamorphic images are pictures that are stretched or condensed so that they appear distorted until viewed from a particular perspective. Some images will be seen on the surface of a cylinder formed from the Mylar sheet tucked in the book's inner cover. Rand's illustrations invite readers to play with the images and stretch their imaginations just as Sandburg's image-laden poem does. Further details for relaying the mathematics of the illustrations and creating similar ones with distorted grids are given on the final pages. Rand's presentation of *Arithmetic* is excellent and should captivate readers of many ages.

Supraner, Robyn. *Stop and Look! Illusions*. Illustrated by Renzo Barto. Mahwah, N.J.: Troll Associates, 1981.

ISBN 0-89375-435-8   (paper) $3.50

★★ Single concept, 1–5

Questions are posed about familiar illusions that involve size comparisons or figures that change, move, or disappear as the reader's perception changes. The book lists readily available materials needed for a variety of interesting experiments. Some of the experiments are making a movie, magnifying objects, and turning objects upside down. Simple step-by-step directions and cartoon characters guide the reader through each experiment. The fascinating experiments, simple text, and colorful drawings should appeal to the young reader.

Time-Life for Children Staff. *From Head to Toe: Body Math.* Alexandria, Va.: Time-Life for Children, 1993.

See Series and Other Resources: I Love Math

Westray, Kathleen. *Picture Puzzler.* New York: Ticknor & Fields Books for Young Readers, 1994.

ISBN 0-395-70130-9 $13.95

☆ Single concept, 1–4

Afterimages and blind spots, impossible drawings, reversible pictures, and trick drawings are among the optical illusions that confuse the brain of the reader. Many of the illusions are classic, but the effective use of color and design has enhanced their appeal. Framed as a painting, one illustration in black and tan shows a vase or two faces depending on whether the viewer is looking at the foreground or background; on the opposite page the colors have been reversed and the images again change. A flower design contains the illusion of two congruent circles that appear to be different sizes; one is surrounded by "petals" of six small circles and the other by "petals" of six large circles. Interpreting two-dimensional drawings of three-dimensional shapes is especially appealing as the reader decides if the book is open or closed, if the butterfly is inside or behind the box, or if the array of colored cubes is being viewed from above or below. A simple, concise explanation is given for each illusion presented in this eye-catching and colorful book of trickery.

White, Laurence B., and **Ray Broekel.** *Optical Illusions.* Illustrated by Anne Canevari Green. New York: Franklin Watts, 1986.

ISBN 0-531-10220-3   (library binding) $11.62

★★★ Single concept, 5–8

"Seeing Is Deceiving," the first chapter of this book, introduces illusions with examples that are not what they seem. Color, movement, and light are some of the factors explored in the perception experiments that are clearly described for the reader's participation. The relationship between what the eye sees and what the brain perceives is fascinatingly presented and keeps the attention of the reader from cover to cover. Observation and experimentation, two sets of skills that form connections between mathematics and science, are encouraged and developed. Illustrations are very clear, adding much to the comprehension and delight of the reader, who will be enthralled by phenomena that are not as they seem. Many intermediate- and middle-grades students will avidly seek access to this book.

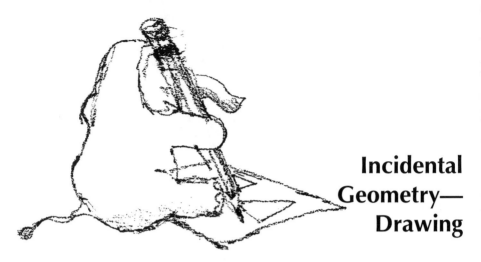

Incidental Geometry— Drawing

Simple geometric shapes, such as circles, rectangles, curved and crooked lines, and so on, are the bases of the completed pictures in these simple drawing books. To follow the step-by-step directions, the reader must distinguish the figure that is added and how it is related to the rest of the drawing.

Emberley, Ed. *Ed Emberley's Big Green Drawing Book.* Boston: Little, Brown & Co., 1979.

ISBN 0-316-23595-4 $15.95

☆ Single concept, K+

Emberley's drawings afford an opportunity to see and compose two-dimensional designs by using open and closed curves. As in Emberley's other how-to-draw books, a simple sketch, such as two intersecting circles or a pair of parallel lines, begins each set of step-by-step diagrams. Underneath each subsequent diagram is a sketch of the shape to be added to the increasingly complex picture. Over a series of four to fifteen diagrams, the two intersecting circles gradually become a picture, such as a woman with a rolling pin or a man with a shovel. Broken and curved lines; intersecting and parallel lines; and shapes, lines, and curves that vary in size and proximity are some of the many discriminations with which the reader interacts to form the final drawings. Eight shapes— triangles, rectangles, circles, semicircles, line segments, arcs, parentheses, and scribbles—are the basic elements of the drawings in this how-to-draw book. Different sections, which vary in complexity, include how to draw people, animals, trees, Dracula, or ships. The drawings and their presentations are often humorous; one section is called "Cat A Log" (how to draw a cat from a log) and another is similarly titled "Dog A Log."

————. *Ed Emberley's Big Orange Drawing Book*. Boston: Little, Brown & Co., 1980.

ISBN 0-316-23418-4 $15.95
ISBN 0-316-23419-2   (paper)   9.95
☆ Single concept, K–5

Witches, spooks, and pumpkins are some of the illustrations of Halloween, Emberley's main theme in this how-to-draw book. Some of the sketches require dividing circles into halves, fourths, and then eighths; another sketch describes how to draw a rectangular prism. The reader is concerned with two-dimensional spatial relationships throughout the book. See *Ed Emberley's Big Green Drawing Book* for the complete annotation of this series.

————. *Ed Emberley's Big Purple Drawing Book*. Boston: Little, Brown & Co., 1981.

ISBN 0-316-23422-2 $15.45
ISBN 0-316-23423-0   (paper)   9.95
☆ Single concept, 1+

Purple pirates, parrots, treasure chests, ships, and other things piratical compose only one of the themes in this how-to-draw book. Emberley includes comments on how to make objects appear smaller or larger by considering their relative size. His sketches of trucks and vans have a three-dimensional perspective; brief comments are included on using a T-square and a triangle as drawing tools. Some final sketches have subtle changes, such as perspective, and the reader is challenged to figure out how to draw the figure. See *Ed Emberley's Big Green Drawing Book* for the complete annotation of this series.

————. *Ed Emberley's Great Thumbprint Drawing Book*. Boston: Little, Brown & Co., 1977.

ISBN 0-316-23613-6 $14.95
ISBN 0-316-23668-3   5.95
☆ Single concept, 1+

Ink, paint, or watercolors can be used to form the imprint of a thumb, which is the basis for all drawings in this book. Sequenced sketches illustrate how to embellish the thumbprint to compose each of the simple drawings. See *Ed Emberley's Big Green Drawing Book* for the complete annotation of this series.

Emberley, Ed, and **Rebecca Emberley.** *Ed Emberley's Big Red Drawing Book*. Boston: Little, Brown & Co., 1987.

ISBN 0-316-23434-6 $14.95
ISBN 0-316-23435-4   (paper)   9.95
☆ Single concept, no age given

American patriotism is the theme of the red, white, and blue section of the book; Christmas is the theme of the red and green section. A variety of animate and inanimate objects can be drawn with the techniques introduced. Estimation is used to divide a line segment into equal parts; grids are introduced to draw objects in equally sized squares. Directions show how to use a compass to transfer equal lengths. Challenges are given to the reader to combine or adapt various drawings. See *Ed Emberley's Big Green Drawing Book* for the complete annotation of this series.

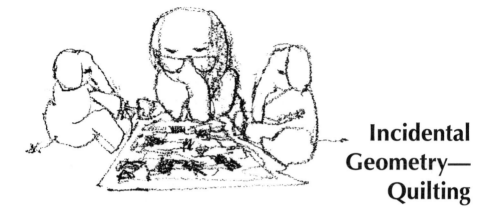

Incidental
Geometry—
Quilting

Quilting has been a recognized art form from the very beginning of America's history. Shared in some variation by virtually every culture to immigrate to America, the quilt has continued to evolve and carry new stories and traditions across the many generations of our constantly changing culture. The role of mathematics in this art form cannot be denied, from the skill involved in precise measurement, to the spatial concepts developed in creating original patterns, to the geometry involved in covering a surface. This section is designed to recognize the powerful mathematics applications found in the art of quilting.

For the purposes of this publication, a sampling of children's literature involving quilting has been included to give the reader an introduction to the many types of literature available for an exploration of quilts. The centrality of the quilt to the story varies, from books clearly designed for constructing quilts to selections where guidance and motivation for quilting are more dependent on the teacher or adult. Three subheadings have been chosen to reflect the importance of quilting and the possible exploration of the mathematics related to quilt construction:

- *Explicit:* Books listed in this category have constructing a quilt as a central focus. This section includes both fiction and nonfiction titles. All selections give some form of direction for the eventual construction of a quilt. The mathematics involved can be specific, such as in the actual measurement and dimensions of a quilt. Other mathematics is more applied, as with the design and pattern of a quilt. Although teacher or adult assistance is important for making a quilt on the basis of most of these selections, some titles, such as *Kids Making Quilts for Kids* or *The Quilt Block History of Pioneer Days* provide enough information for older children to work independently on a quilting project.

- *Implicit:* Titles in this section are fiction, but many have a historical basis or are retellings of family history. The construction of a quilt is central to the story line in these books, and the process of quilt making is often fairly clear. The reader receives a strong message that quilts have a varied and important role in our family and cultural traditions. Geometry is integral, since the reader is strongly encouraged to problem solve by recreating specific quilt-block patterns. Some selections, such as *Jumping the Broom,* include a specific embedded-mathematics problem as readers are invited to wonder how many diamond shapes will be needed to construct an eight-pointed-star quilt given that a total of seventy-two diamonds is needed for each block. Although the invitation to explore mathematics is strong, guidance from an adult remains vital to the development of good mathematical connections.

- *Incidental:* These stories are fiction or are retellings of family stories. All have the quilt as a central theme. However, the actual process of quilt making is given a brief role in the overall story. Connections or springboards to the construction or design of an actual quilt depend on adult interaction with the young reader. Mathematics connections would need to evolve from whatever experiences were designed from the use of these books. These books are included in recognition of the importance of integrating mathematics into all curricular areas and vice versa. Rich mathematical connections are often made in the context of the classroom sharing of literature in this category and are frequently the result of extensions contrived by the reader.

Explicit

ABC Quilts. *Kids Making Quilts for Kids.* Gualala, Calif.: The Quilt Digest Press, 1992.

ISBN 0-913326-36-0   (paper) $9.95

★★★ Multiconcept, 4–8

This informational text provides the reader with specific information about constructing quilts. Enhanced by photographs of actual quilts designed and sewn by groups of students in a school setting, this book gives clear, step-by-step instructions for every step from selecting the right fabric to using metric conversion charts. This resource would be invaluable in launching a quilt project regardless of the level of sophistication. Black-and-white diagrams supplement the color photos and provide basic patterns for a number of beginning quilt designs. Although this book was the result of student-service projects dealing with substance abuse and AIDS, it offers useful tips for creating quilts for any purpose.

Bolton, Janet. *My Grandmother's Patchwork Quilt: A Book and Pocketful of Patchwork Pieces.* New York: Bantam Doubleday Dell Publishing Group, 1994.
ISBN 0-385-31155-9 $17.95
★★★ Multiconcept, K–5

The author clearly intends readers to try their hand at appliqué techniques to create a unique doll's quilt. Told from the viewpoint of a grandchild, this story describes how a child recreated her grandmother's patchwork quilt. The quilt, which is an account of her grandmother's daily life on the farm, is shown patch by patch on the book's sturdy cardboard pages. Each patch is described from two points of view. On the left page, the author gives a third-person account of the grandmother's life. On the right side, the author presents a quotation from the grandmother that tells the significance of the finished quilt block. As each block is described, it is shown in the context of the overall quilt pattern. By the book's end, the entire quilt has been reconstructed, and the reader has been given a step-by-step guide to making this quilt. At the back of the book, the author provides the quilt patches needed to construct this quilt. Color photography of the actual quilt blocks by Bolton, a fabric artist, adds a unique dimension to this book.

———. *Mrs. Noah's Patchwork Quilt: A Journal of the Voyage with a Pocketful of Patchwork Pieces.* Kansas City: Andrews and McMeel, 1995.
ISBN 0-8362-4250-5 $17.95
★★ Multiconcept, K–5

Janet Bolton uses her skill as a fabric artist to replicate the steps involved in the creation of Mrs. Noah's patchwork quilt. Similar in format to her earlier publication, *My Grandmother's Patchwork Quilt,* this book tells the story of the forty days and forty nights Mrs. Noah spent at sea on the ark with all the animals. The left-hand page tells the story of Noah's ark from a third-person perspective; the right-hand side tells the story from the first-person perspective of Mrs. Noah. Again, internationally known textile artist Janet Bolton treats the reader with a photograph of each quilt block as the quilt takes form throughout the book's pages. Although the subject matter of this book might appeal to only a select audience, the step-by-step approach to the construction of an appliqué quilt would interest many groups. Again, fabric patches are included in a pocket at the back of the book, a direct invitation for the reader to recreate Mrs. Noah's patchwork quilt.

Cobb, Mary. *The Quilt-Block History of Pioneer Days: With Projects Kids Can Make.* Illustrated by Jan Davey Ellis. Brookfield, Conn.: Millbrook Press, 1995.

ISBN 1-56294-485-1   (library binding) $17.40
ISBN 1-56294-692-7   (paper)   7.95

★★★ Multiconcept, 2–6

This delightful introduction to historical quilt patterns is associated with the pioneers' movement west from the 1700s to the 1800s in the United States. Each section of this book deals with a different aspect of the journey west, both to give the reader some perspective on that time in history and to present the functional aspect of the quilt patterns. At the book's outset, the author provides the reader with a simple history of the role of the quilt in our nation's development. From there, each chapter focuses on a particular event or set of events typical on any journey west. Along with each description is an accompanying quilt pattern that arose from specific needs of the pioneer families making their difficult journey. Each section describes and illustrates at least four historical quilt-block patterns. Some create a border for the text on the page, whereas others hang from an old-fashioned clothesline. In addition to clear illustrations that invite the reader to replicate the pattern in some way, each section describes a specific project, based on a quilt pattern, that the reader can construct with readily available materials. A bibiliography of further reading about pioneer travel is included at the end of this clearly written account of pioneer quilt patterns.

Emberley, Ed. *Ed Emberley's Picture Pie: A Circle Drawing Book.* Boston: Little, Brown & Co., 1984.

See Geometry and Spatial Sense: Geometric Concepts and Properties

Friedman, Aileen. *A Cloak for the Dreamer.* Illustrated by Kim Howard. New York: Scholastic, 1994.

See Geometry and Spatial Sense: Geometric Concepts and Properties

Kurtz, Shirley. *The Boy and the Quilt.* Illustrated by Cheryl Benner. Intercourse, Pa.: Good Books, 1991.

ISBN 1-56148-009-6   (paper) $6.95

★★ Multiconcept, K–3

This humorous story, with its slightly zany illustrations, tells of a young boy who decides to make a quilt on his own. The book is easy to read and may deceive the reader at first glance. The story actually gives a very clear description of the steps necessary for constructing a basic quilt, from carefully measuring and cutting the squares to actually quilting the layers. A warm, family-centered story, it depicts the many considerations that go into making a quilt. An added feature is the depiction of a boy in the role of an interested quilter. Another, very helpful feature is the inclusion of a two-page set of directions for making a quilt similar to the one in the book. Beginning quilters will appreciate the author's description of how to make a patchwork piece into a tied comforter instead of an actual quilt.

Paul, Ann Whitford. *Eight Hands Round: A Patchwork Alphabet.* Illustrated by Jeanette Winter. New York: Harper Collins, 1991.

ISBN 0-06-024689-8		$15.00
ISBN 0-06-024704-5	(library binding)	14.89
ISBN 0-06-443464-8	(paper)	4.95

★★★ Multiconcept, 2–5

An excellent historical narrative, this book shares the significance of many authentic patchwork patterns. The clear text should appeal to young readers interested in the history of the United States and in the stories behind many familiar quilt patterns. Although step-by- step instructions for constructing the illustrated quilt blocks are not included, the patterns are easily replicated by using various combinations of squares, right triangles, and isosceles triangles. Whereas a young historian would enjoy these short descriptions, a young quilter would benefit from adult assistance when attempting to recreate the patterns included in the book. Opportunities to explore geometric patterns and spatial concepts abound in this creative and informative alphabet book enhanced by the earthy, colorful illustrations of Jeanette Winter.

———. *The Seasons Sewn: A Year in Patchwork.* Illustrated by Michael McCurdy. New York: Harcourt Brace & Co., 1996.

ISBN 0-15-276918-8	(hardcover)	$16.00

★★★ Multiconcept, 3–8

Paul follows her popular *Eight Hands Round* with a collection of more intricate patchwork patterns that have evolved from stories of American history. This book's increased sophistication is enhanced by McCurdy's appealing scratchwork illustrations. The patchwork patterns are clearly illustrated in bright colors alongside the brief historical description. Although specific instructions for constructing these quilt blocks are not given, the reader is invited to discover the way in which the patterns are made. Reminiscent of Paul's earlier work, the patterns consist primarily of designs built from various-sized squares and triangles but with a higher degree of complexity. Readers more experienced with geometric patterns and spatial relationships will be able to recreate these patterns. Organized around the four seasons, this book is an outstanding springboard into patchwork art.

Time-Life for Children Staff. *The House That Math Built: House Math.*
Alexandria, Va.: Time-Life for Children, 1993.
See Series and Other Resources: I Love Math

Implicit

Ernst, Lisa Campbell. *Sam Johnson and the Blue Ribbon Quilt.* New
York: Lothrop, Lee & Shepard Books, 1983.
ISBN 0-688-11505-5   (paper) $3.95
☆ Multiconcept, 1–4

Young readers will enjoy this old-fashioned story involving role
reversal. Sam Johnson sees the value of using thread and needle on
the farm and becomes naturally interested in the art of quilting.
Unable to convince his wife to allow him to join her women-only
quilting club, Sam forms his own. Composed only of men, his quilting
circle designs a quilt to enter into a competition with the women at
the fair. As luck would have it, the two groups accidentally collide
enroute to the fair, which destroys their individual quilts. In the
spirit of cooperation, the two groups problem solve to blend the two
quilts into a new quilt with a new pattern. As might be predicted, the
combined efforts of the men and women produce a blue-ribbon quilt.
Actual patchwork quilt patterns form the border of each page, and
each pattern relates to the content of the story on that page. Easy to
read, this book invites an exploration of pattern and design. Pen-and-
ink drawings with muted color create illustrations that not only are
humorous but also are reminiscent of earlier times. Readers are
introduced to the notions that more than one pattern can be produced
from the same quilt pieces and that quilt construction requires
careful measuring and craftsmanship.

Flournoy, Valerie. *The Patchwork Quilt.* Illustrated by Jerry Pinkney.
New York: Penguin Books USA, 1985.
ISBN 0-8037-0097-0 $15.99
ISBN 0-8037-0098-9   (library binding)   14.89
★★★ Multiconcept, 1–3

This tender story illustrates the significance of the patchwork
quilt in the preservation of family memories and traditions. Told
within the framework of one full calendar year, this story develops
concepts of time as a young girl begins to appreciate family tradition
and quilting by helping her grandmother construct a family memory
quilt. Pencil, graphite, and watercolor illustrations by award-winning
artist Jerry Pinkney bring this story to life. Based on events of an
African American family, this recipient of the Coretta Scott King
Award transcends all cultures. Selected as a Reading Rainbow Book,
The Patchwork Quilt encourages direct investigation into basic
quilting. Time concepts for one year can also be developed with the
framework of this story.

Hopkinson, Deborah. *Sweet Clara and the Freedom Quilt.* Paintings by
James Ransome. New York: Alfred A. Knopf, 1993.

ISBN 0-678-82311-5		$16.00
ISBN 0-679-92311-X	(library binding)	16.99
ISBN 0-679-87472-0	(paper)	6.99

★★★ Multiconcept, 2–5

Colorful, textured paintings by James Ransome illustrate the
retelling of the story of a young slave girl who is inspired to show the
way to freedom by constructing a quilt. Modeled after the experiences
of slaves who actually lived on Verona Plantation, this historical
story promotes an understanding of the Underground Railroad and
the significance of the quilt for marking routes to freedom in the
pre–Civil War era. In addition to teaching historical concepts, this
story is a natural springboard for the development of spatial concepts
through mapmaking. The paperback edition specifically invites
readers to consider an exploration of mapmaking and various quilt
patterns of the period. This story strongly suggests an investigation
into number, geometry, and patterns through quilt construction. This
book has been identified as a Reading Rainbow Book and a
Children's Book-of-the-Month Club Selection and has received the
International Reading Association Award.

Ringgold, Faith. *Dinner at Aunt Connie's House.* New York: Hyperion
Books for Children, 1993.

ISBN 1-56282-425-2		$15.95
ISBN 1-56282-426-0	(library binding)	14.89
ISBN 0-78681150-1	(paper)	4.95

★★ Multiconcept, 2–5

Based on Ringgold's painted story quilt, *The Dinner Quilt,* this
story tells of a special visit by two children to Aunt Connie's for
Sunday dinner. During the course of the visit, the children discover a
special surprise: Aunt Connie's paintings that magically tell their
own stories. Each painting is of an African American woman who
struggled to make a special contribution to the United States. The
children's discovery is shared, and they proudly help display Aunt
Connie's works of art around the Sunday dinner table for all to enjoy.
This book shows how art can be more than a picture on a wall—it can
enliven history and magically illustrate proud events in people's
lives. Paint-on-canvas illustrations are done by Ringgold. Also
included is a reproduction of Ringgold's *The Dinner Quilt,* which can
be used to entice children to create a story quilt of their own.

Smucker, Barbara. *Selina and the Bear Paw Quilt.* Illustrated by Janet
Wilson. New York: Crown Publishers, 1995.

ISBN 0-517-70904-X		$16.00
ISBN 0-517-70910-4	(library binding)	17.99

★★★ Multiconcept, 2–5

Young readers expand their understanding of the different groups who chose not to remain in the United States during the Civil War. In this well-written story, a Mennonite family faces both the conflict of the Civil War and the personal conflict involved in avoiding religious persecution. Rather than oppose the teachings of their Mennonite faith, these family members choose to flee their home in Pennsylvania to start a new life in Canada. Too old to travel with the family, Selina's grandmother remains behind. A special patchwork quilt given as a gift to Selina provides a loving link between both generations and both countries. Photographs of actual quilting patterns form the borders of the painted illustrations on each page. The bear-paw pattern and the log-cabin pattern are featured in both the text and the illustrations. Many invitations to construct meaningful quilt-block patterns exist within the pages of this authentic story.

Wright, Courtni C. *Jumping the Broom*. Illustrated by Gershom Griffith. New York: Holiday House, 1994.

ISBN 0-8234-1042-0 $15.95

★★★ Multiconcept, 2–5

A story of the marriage of two slaves develops cultural appreciation for the customs surrounding a wedding celebration in the pre–Civil War era. The women in the slave quarters work at night to create a special "star quilt with eight points on each star." The creation of the star quilt invites an investigation of number and patterning with the description that each eight-pointed star was made "by stitching together seventy-two small diamond pieces." Geometric and spatial concepts are strongly implied through the skills involved in creating a star from diamond shapes. Gershom Griffith's bright watercolor illustrations give strong visual clues for the star pattern, including the prescribed colors, "three shades of blue in every star," and "many reds, yellows, and greens."

Incidental

Howard, Ellen. *The Log Cabin Quilt*. Illustrated by Ronald Himler. New York: Holiday House, 1996.

ISBN 0-8234-1247-4 $15.95

★★ Multiconcept, K–3

Beautiful paintings by acclaimed children's book illustrator Ronald Himler adorn the text of this touching story of struggle and hardships in pioneer life. The story provides the young reader with an appreciation of the role of the patchwork quilt in pioneer life. It also affords the young reader an opportunity to investigate the double meaning of the term *log-cabin quilt*. Usually associated with a specific geometric pattern formed by strips of varying lengths, the term *log-cabin quilt* takes on its literal interpretation as quilt scraps are used to fill in the cracks of the newly constructed log cabin during the winter season. Readers may be motivated to explore the geometry inherent in the authentic log cabin pattern after reading this story.

Johnston, Tony. *The Quilt Story.* Illustrated by Tomie dePaola. New York: G. P. Putnam's Sons, 1985.

ISBN 0-399-21009-1		$15.95
ISBN 0-698-11368-3	(paper)	5.95
ISBN 0-398-22403-3	(paper)	5.95

☆ Multiconcept, PS-3 (K–2)

Young fans of beloved children's author-illustrator Tomie dePaola will recognize his artistic style in the pages of this quilt story. The significance of the quilt in providing comfort and building family memories is inherent in this pioneer story. The timelessness of the art of quilting is also evident as the story and the quilt span time from colonial days until modern times when the old quilt is rediscovered. The age-old interest in quilts is highlighted in this story as a mother and child restore an old family quilt. Opportunities for exploring or constructing quilts and their many patterns are highly dependent on adult leadership.

Polacco, Patricia. *The Keeping Quilt.* New York: Simon & Schuster, 1988.

ISBN 0-671-64963-9	$14.95

★★★ Multiconcept, K–3

Artfully told and illustrated by Patricia Polacco, this heartwarming story comes directly from her family history. An account of her family's tradition of passing down its memory quilt for almost a century, this story illustrates the significance of the quilt in preserving family history and its universal appeal for many cultures. Children are invited to appreciate the many adaptations a quilt can undergo in its life, incorporating new and meaningful memories as it evolves and is passed from generation to generation. This book received the Sydney Taylor Award from the Association of Jewish Libraries. Extremely effective are the charcoal drawings splashed with color to highlight the significant articles of clothing that eventually make up the family's keeping quilt. Patricia Polacco's style as an author and an illustrator is highly evident in this invitation to quilting.

Ringgold, Faith. *Aunt Harriet's Underground Railroad in the Sky.* New York: Crown Books for Young Readers, 1993.

ISBN 0-517-58767-X		$15.00
ISBN 0-517-48768-8	(library binding)	17.99
ISBN 0-590-47781-1	(paper, Scholastic)	4.95
ISBN 0-517-88543-3	(Crown Publishers)	5.99

★★ Multiconcept, 1–3

A young girl and her brother take a fantasy trip on the Underground Railroad. Led by the spirit of Harriet Tubman, the children develop an understanding of the Underground Railroad. In addition, the significance of the star-quilt pattern in our country's pre–Civil War history is introduced. This story affords opportunities to guide the development of geometric thinking involved in star patterns found on historical quilts. The design for a quilt commemorating Harriet Tubman and the 100th anniversary of the Underground Railroad is depicted at the story's end. The informational text following the story presents a short biographical sketch of Harriet Tubman and her work with the Underground Railroad. Author and world-reknowned artist and quilter Faith Ringgold illustrates the story with original paintings.

Ross, Kent, and **Alice Ross.** *Cemetery Quilt.* Illustrated by Rosanne Kaloustian. Boston: Houghton Mifflin Co., 1995.

ISBN 0-395-70948-2 $14.95

☆ Multiconcept, K–3

Confronting the sensitive subject of the death of a family member, this story provides the young reader with yet another role that quilts play in our varied culture. A young girl deals with the death of her grandfather as she is introduced to the tradition of adding names to the symbolic cemetery quilt to remember family members who have died. Award-winning artist Rosanne Kaloustian provides the paintings that set the mood for this touching story of family love, remembrance, and tradition. Recommended for specialized audiences and purposes, this story illustrates the powerful role of quilts in our society.

Wright, Courtni C. *Journey to Freedom: A Story of the Underground Railroad.* Illustrated by Gershom Griffith. New York: Holiday House, 1994.

ISBN 0-8234-1096-X   (library binding) $15.95
ISBN 0-8234-1042-0   (library binding)   15.95

★★ Multiconcept, 2–4

Author Courtni Wright teams up once more with Gershom Griffith to tell a story set in pre–Civil War United States. A young boy and his family are led by Harriet Tubman from slavery in Kentucky to freedom in Canada on the Underground Railroad. Griffith's realistic paintings depict the final days of the family's journey to freedom, which are frought with danger and discomfort. Once again Wright introduces a significant use of the quilt in this period of our country's history. Included in this story are references not only to the quilt as a signal for safe houses on the route to freedom but also to the significance of the color black in the signal quilt. The inclusion of black meant that travelers on the Underground Railroad could walk into the yard because "all's clear." A simple quilt pattern constructed from black, blue, crimson, purple, and pink rhombus shapes would be a meaningful application of tessellating regular polygons. Readers may note the varied designs of quilts used along the Underground Railroad and do further explorations with tessellating patterns for authentic quilt blocks.

**Series and
Other Resources**

S ue Weinberg and Cassi Heintz at Dike Elementary School decided to use one of the stories from Time-Life's *Alice in Numberland* in their exploration of fractions with their second- and third-grade students. An active discussion evolved after the initial reading of "A Piece of Cake." The story line involved cutting a circular cake into halves, thirds, fourths, and sixths. The children could identify with cutting and decorating birthday cakes. Questions from the text were addressed, for example, whether halves or thirds were larger or how many of the cake's twelve decorative roses would be on each piece.

Over the next few days this story was the impetus for folding fractional circles and identifying the parts. Equivalent parts, related fractions, equivalent fractions, and other fraction concepts and appropriate vocabulary evolved naturally. The class quickly discovered the pattern for folding halves, fourths, eighths, and sixteenths. Joel, a third grader, showed his classmates how to fold sixths by folding halves and then thirds. The third graders confidently discussed and wrote other names for one-half, such as two-fourths, three-sixths, and four-eighths.

For one activity, each student decorated an individual "cake" and decided into how many parts to divide the cake and how many rose decorations should be on each slice. Emily, a second grader, folded a paper circle into fourths, drew three roses on each slice, and counted by multiples of 3 to arrive at twelve roses. On a second problem, she divided the cake into sixths, drew two roses on each slice, and determined the number of roses by counting by multiples of 2.

Second graders Colin and Libby were working on sharing ten roses among six people. They drew ten circles for ten roses and six circles for six people and then drew lines indicating that each person would get one whole rose.

Next, they cut the next three roses into halves and drew a line indicating a half for each person. The final rose was then cut into six parts. Colin and Libby could see that half a rose was the same as 3 sixths of a rose, so 1/2, or 3/6, plus 1/6 more would be 4/6.

The description above highlights one book from the I Love Math series. A valuable resource for both Carmen DeVoe and her fourth graders has been Vorderman's *How Math Works.* Carmen has used it as a resource for her classroom and her work in a master's degree program. Her fourth graders actively seek it out when they want more information about specific mathematics topics. *How Math Works,* a resource, and *Marvelous Math,* an anthology of poems, have been placed in this category because they are unique entities. The remainder of this category comprises books that belong to a series.

In the first edition of *The Wonderful World of Mathematics,* books within series were referenced in the appropriate content section. For this edition, a new section was created as the result of dilemmas encountered in reviewing series.

- Books in a series contain common elements, which were often presented only in the first review. The reader of subsequent volumes was referred back to the first review. The remainder of a review would highlight the content and features specific to that book.
- If the series contained a number of topics, like those written by Pluckrose, books were reviewed in many different sections and the reader did not gain a perspective of the entire series.

For these reasons and the expanded availability of book series, a separate section titled Series and Other Resources has been created. Instead of being alphabetized by author, the individual series were alphabetized by title and all books listed under that series were then alphabetized by author. An annotation for the overall series is followed by annotations of the individual books in the series. Each book is also cross-referenced in appropriate content sections. The following series were reviewed in this format:

Action Math
Discovering Math
Discovering Shapes
First Step Math
Let's Investigate
Math Counts
MathStart
Measure Up with Science

Two series were placed entirely in a content section because each addressed a single topic. Reviews for Haskin's Counting Your Way can be found in the Counting in Other Cultures section, and Morgan's The World of Shapes was classified in Geometry and Spatial Sense: Shapes.

Resources

Marvelous Math: A Book of Poems. Poems selected by Lee Bennett
Hopkins. Illustrated by Karen Barbour. New York: Simon & Schuster
Books for Young Readers, 1997.
ISBN 0-689-80658-2 $17.00

★★ Single concept, K+

Brilliant artwork provides a colorful setting for the sixteen poems
about mathematics written by, or selected by, Hopkins. Barbour's
vivid colors and flat style are reminiscent of Gauguin; the artwork
was composed of layers of gouache paint with figures outlined in
black on a vibrant background. Poems like O'Neill's "Take a Number"
focus on how we use mathematics in our everyday lives and what our
world would be like without number and measure. Some poems are
playful, like McCord's account of division, titled "Who Hasn't Played
Gazintas?" Brevity is a prominent feature of Dotlich's "Calculator": in
four lines she describes this tool and concludes the line with "Mini-
magician." Some poems depict a stereotyped view of mathematics
that is relegated to arithmetic and its avoidance; some poems
incidentally refer to mathematics. In spite of these limitations, this
anthology is a helpful addition to children's literature in
mathematics and can motivate readers to explore other sources or
write their own poetry.

Vorderman, Carol. *How Math Works.* Pleasantville, N.Y.: Reader's
Digest Association, 1996.
ISBN 0-89577-850-5 $24.00

★★★ Multiple concept, 3–9

Published for the Reader's Digest Association, this 192-page, 8 1/2-
inch-by-11-inch resource is another quality production by the project
staff at Dorling Kindersley. Designed for students from ages 8 to 14,
How Math Works is part of a series to help them appreciate
and understand the world of scientific discoveries. In this volume,
mathematics is organized into seven chapters: "Numbers," "Prop-

ortions," "Algebra," "Statistics," "Measurement," "Shape," and "Thinking." Numerous subtopics are included in each chapter; some of the subtopics for number include a history of writing numbers, series and sequences, calculating tools, positive and negative numbers, fractions and decimals, estimation, powers, and number bases. Each two-page spread, highlighting a subtopic, is filled with information and activities, but it is effectively organized and presented with an attractive layout incorporating color, diagrams, photographs, partitions, and different sizes of font. For example, the title of each subsection and a brief description are printed in larger fonts. Discoveries by famous mathematicians are framed by a rectangle with a buff-colored background. Historical artifacts like Egyptian papyrus are pictured and described and set off in rectangular frames or spaces. Experiments and activities are set off by headers and larger fonts. A "You Will Need" legend lists and pictures the necessary materials; numerals guide the reader through the step-by-step directions. As an example of the scope and substance of this resource, some of the information found in one section includes discoveries by Archimedes, John Napier, Shakuntala Devi, Isaac Newton, Maria Goeppert Mayer, Gottfried Wilhelm Leibniz, and Michael Faraday; the origins of triskaidekaphobia, a super-stitious fear of the number thirteen; square and cubic numbers; the history of the mechanical calculator; the Wall Street crash of 1929; and negative altitudes. Some of the experiments, demonstrations, and investigations include adding numbers on a Chinese abacus, halving and doubling, casting out nines, making Napier's bones, exploring divisibility tests, using pulleys, calculating with zero, and understanding powers. Puzzles and tricks include magic squares, calculator tricks, and factors. These partial listings are representative of the entire book; fascinating facts to uncover and interesting investigations to actively involve the reader are also included in the other chapters. A glossary, answers to selected problems, and a comprehensive index of topics and mathematics are included. This well-designed resource is highly recommended for families, classrooms, and libraries.

Action Math

All the books in the Action Math series have a similar format and include close-up, color photographs of the various projects. Very explicit step-by-step directions for making the projects are printed in large type. Young readers will need adult assistance in assembling and constructing the projects; paint, scissors, glue, and cardboard are often needed. Occasionally a mismatch occurs between the age of the child who could construct the projects and the age of a child who would learn from the project. Overall the projects are of high quality; some would appeal to preschoolers and others would interest fourth graders. The purpose of each task is highlighted in a gray rectangle captioned "Here's what you learn." Features in each book also include a table of contents and a "Skills Index," which lists mathematical purposes. Wendy and David Clemson were consultants to the series.

Bulloch, Ivan. *Games.* New York: Thomson Learning, 1994.
ISBN 1-56847-231-5 $14.95
★★ Multiconcept, (K–3)

Games includes card and board games that involve a variety of mathematical skills, such as recognizing patterns, numbers, and shapes and matching and ordering numbers. Color photographs and clear instructions illustrate how to construct the game pieces as well as how to play each game. Card games like rummy and snap use simple number concepts. Board games and bingo form the basis for games that deal with either geometric or number patterns. Probability concepts are embedded in the beetle game as each player tosses a die to determine whether an antenna, a head, a tail, and so on, can be added to the beetle's body. Unique versions of jigsaw puzzles and tic-tac-toe are also described and illustrated through color photographs.

————. *Measure.* New York: Thomson Learning, 1994.

ISBN 1-56847-233-1 $14.95

★★ Single concept, (1–3)

Measure includes several projects for comparing sizes and using measures to explore lengths, heights, weights, areas, and volumes. Some projects have students make a chart to measure their height, investigate size by making hand and finger puppets, measure to create exotic face masks, match sizes and shapes by constructing paper dolls, and explore area and shape with tangrams. A project that uses measurements to bake cookies is cleverly followed by volume and area explorations; boxes are made to package the cookies, and the amount of colorful paper needed to wrap the boxes is estimated.

————. *Patterns.* New York: Thomson Learning, 1994.

ISBN 1-56847-230-7 $14.95

★★★ Single concept, (1–3)

Patterns primarily centers on geometric patterns. On the opening pages, color photographs highlight zebra stripes, leaves, an orange's cross section, and other interesting patterns in the world around us. The reader is invited to participate in a number of illustrated projects. One of the projects teaches weaving; patterns extend into diagonal, wavy, tartan, and zigzag stripes. Another project has readers cut paper to form symmetrical patterns. Symmetry is also explored as readers fold freshly painted paper to create unique designs. Congruent squares colored with the same two-color design are arranged and rearranged to form new patterns. Mosaics are created with tiles of rhombuses, squares, rectangles, and triangles. Other projects include cutting spirals to form a snake and constructing and threading beads to form a pattern.

————. *Shapes.* New York: Thomson Learning, 1994.

ISBN 1-56847-232-3 $14.95

★★★ Single concept, (1–3)

Shapes centers on two- and three-dimensional shapes. Color photographs on the opening pages highlight shapes in the world around us: ice-cream cones, a die, a plate, and coins. One two-dimensional project includes folding and cutting paper to form squares, triangles, and semicircles or to explore designs having one or two lines of symmetry. In another project readers fit together hexagons to form a tiling pattern. "Slits and slots" enable two-dimensional cardboard shapes to be cleverly placed together to form three-dimensional figures. These ideas are extended to build entire scenes, such as jungles that use cardboard-cylinder tubes for tree trunks. Patterns for constructing cylinders, cubes, and triangular and rectangular prisms are given. Readers are invited to make collages by using only familiar two-dimensional shapes or only three-dimensional shapes.

Discovering Math

Number and operations are the focus of the six-volume set of the Discovering Math series. Each thirty-two-page book is a collection of activities introduced in a one- or two-page format. Primarily it is a collection for practicing skills, not introducing concepts. Each opening activity presents and defines the topic that is the theme of the book. A wide range of skill levels are incorporated throughout each book; the reader is assumed to have a background in these topics. Often the skills practice is posed in a game, riddle, or puzzle format. As readers successfully solve different computations or problems, they are able to find their way through mazes, decipher codes, or reach the goal at the end of a board game. Number tricks appear throughout and should prove highly motivating. They are presented as neat tricks to amaze others, but no explanation is given for the reasoning behind the task. Activities are often extended by asking readers to create or find more examples on their own. Often such materials as calculators, dice, playing cards, or spinners are needed, or a diagram from the book must be traced and cut out. Connections to history and real life are included when appropriate. Definitions, additional directions, or other information is often conveyed by "balloon speech" from cartoon characters. The illustrations are simply done and are often restricted to diagrams. Each book contains a table of contents, an answers section, a glossary, and an index. Overall, the presentation is uneven; there are excellent, well-developed activities and games as well as contrived ones that look more like those in a traditional textbook.

Stienecker, David L. *Addition.* Illustrated by Richard Maccabe. New York: Benchmark Books, 1996.

ISBN 0-7614-0593-3 $12.95

☆ Single concept, 3–5

Activities from introducing the concept to using the algorithm with three-digit numbers are included in *Addition*. An opening activity is a story situation in which the reader is asked to determine the number of geese in two flocks by drawing a picture of, and then counting, all the geese. Basic-fact knowledge, as well as logical reasoning, is needed to solve various number puzzles in which the addends are to be arranged in a triangular shape or a magic square. Strategy and mental computation are involved in the target addition game that uses a calculator's memory. For a classic number trick, a mystery number can be guessed if the solver knows on which lists of numbers it occurs; another interesting trick involves marking off a four-by-four array of numbers on a calendar and using a shortcut to find the sums of the numbers. Playing cards and dice are needed for a game to find the greatest sum; this is an excellent place-value game. Applications to geometry are incorporated by measuring and adding lengths to find the perimeter of different shapes formed by puzzle pieces of two congruent right triangles. Other perimeter problems are embedded in a maze activity. Palindromes are also introduced; a final addition magic square uses decimals.

————. *Division*. Illustrated by Richard Maccabe. New York: Benchmark Books, 1996.
ISBN 0-7614-0596-8 $12.95
☆ Single concept, 3–5

Determining the number of quilt blocks that can be made from fifteen triangular pieces is the opening activity to introduce division. Basic facts are practiced in a bingolike game as well as with a board game. A calculator is used to explore the relationship between division and subtraction; it is also used to explore dividing by multiples of ten. Number tricks are presented for determining whether a number is divisible by 9 or 3. Problems involving remainders are embedded in activities with magic squares, shading grids, and mazes. A cross-number puzzle is the vehicle for practicing division with one- and two-digit divisors. Twenty-six long-division problems need to be answered to solve "Division Decode." Logical reasoning and an understanding of the division algorithm are essential for successfully solving the division problems that have some of the calculations missing. Like other books in this series, this one describes an excellent game involving place value and the greatest quotient. An activity like the division problems but presented with Mayan symbols is a contrived format, but problems like "Honey Pots" are excellent. A misprint in a problem dealing with deciphering codes should be corrected to 509 divided by 7 rather than 508.

————. *Fractions*. Illustrated by Richard Maccabe. New York: Benchmark Books, 1996.
ISBN 0-7614-0598-4 $12.95
☆ Single concept, 3–5

Determining the fractional amounts of different colors for flags that represent each letter of the alphabet is the opening activity. Dividing various shapes, such as a rectangle, a triangle, a circle, a hexagon, and the letter E, into specific fractional parts is a well-designed and challenging task for which multiple solutions are possible. Tangram pieces are also used to explore fractions by having the reader identify the value of each piece. One card game is designed for practicing ordering fractions; a concentration-like game is designed for identifying equivalent fractions. As in *Addition,* magic squares and other geometric shapes are used as game boards to practice addition, but in this situation, the addition of fractions is included. Subtraction, multiplication, and division are embedded in other games, puzzles, and mazes throughout the book. One particularly good task is to determine two fractions when their sum and their product are given. Other activities include graphing fractions as ordered pairs to form a picture on a grid, using a calculator to change fractions to decimals, and exploring decimal patterns of fractions like one-ninth, two-ninths, three-ninths, and so on.

———. *Multiplication.* Illustrated by Richard Maccabe. New York: Benchmark Books, 1996.

ISBN 0-7614-0595-X $12.95

☆ Single concept, 3–5

If a Ferris wheel has sixteen cabs that can each seat three people, what is the maximum number of people who can ride at once? This situation is the context for opening a discussion on multiplication. A much larger number is the result when Mr. Ferris's 265-foot-high big wheel from the Chicago World's Columbian Exposition is considered; each of the thirty-six passenger cabs could hold forty people! Basic facts are the basis of another activity; the facts are arranged in a table, and as particular facts are shaded, the answer to a riddle appears. Like other books in this series, this one presents multiplication problems in calculator word games and deciphering alphabet codes. Like in *Addition,* a number trick involving numbers in any three-by-three array of calendar squares is posed, but this trick uses multiplication. Some activities involve the multiplication patterns for 9 and 11. After some of the simpler patterns with 9 have been introduced, others, such as $9 \times 9 + 7 =$, $98 \times 9 + 6 =$, $987 \times 9 + 5 =$, ..., $98\,765\,432 \times 9 + 0 =$, are explored. Like *Addition,* this book makes connections to geometry by having readers combine various puzzle shapes to form a rectangle. The reader is then to determine the rectangle's area by multiplying the measure of its length by the measure of its width. Diagrams of mobiles afford another multiplication opportunity when the formula for balancing objects is presented as a relationship between length and weight. Once again, a greatest-product game is described; this excellent game incorporates multiplication and place-value concepts. Factor trees are introduced, but no examples illustrate the same result in more than one way. The power of doubling is shown with a table giving the results of earning $0.01 the first day, $0.02 the second day, and $0.04 the third day; the problem is to determine the amount at the end of the month. A misprint on a problem, "Find the Missing Digits," simply repeats the same problem twice.

————. *Numbers.* Illustrated by Richard Maccabe. New York: Benchmark Books, 1996.
ISBN 0-7614-0597-6 $12.95
☆ Single concept, 3–5

Becoming acquainted with other numeration systems by interpreting the values of numbers written with Egyptian, Babylonian, Roman, Greek, and Mayan symbols is the purpose of the first activity in *Numbers.* Binary numbers are also explored in one activity. One place-value strategy game uses six digits and place values to hundred thousands to find the smallest number. In addition to numeration, different types of numbers are included: diagrams present triangular and square numbers; a maze can be solved by distinguishing between even and odd numbers; the sieve of Eratosthenes is used to find prime numbers. Characteristics of different numbers can be explored in the game "guess the number?" Number pairs on a grid are used for drawing pictures. Names for big numbers like *googol* (1 followed by 100 zeroes) or *vigintillion* (1 followed by 63 zeroes) are presented as well as interesting facts about large numbers, such as our universe has 200 000 000 000 000 000 000 stars.

Wells, Alison. *Subtraction.* Illustrated by Richard Maccabe. New York: Benchmark Books, 1996.
ISBN 0-7614-0594-1 $12.95
☆ Single concept, 3–5

From introducing the subtraction concept to subtracting multidigit numbers, a wide range of activities is described. Opening activities involve both take-away situations (tossing popcorn pieces into a container) and comparison situations (the number of red versus the number of black checkers on a board). Data collection and tallies are involved in one activity similar to a task in *Addition.* Pasta pieces are used like base-ten materials to represent and subtract two-digit numbers; however, since both numbers are represented, young children might find the activity confusing. An excellent task incorporating logical reasoning, subtraction, and ordering has readers compare the speed of a human with those of nine different animals. Another good task involves estimating the heights of several objects that when stacked would form a combined height close to sixty inches; the goal is to minimize the difference between the combined height and sixty inches. A similar task involves a shopping spree to choose from sixteen different items a subset whose total cost would be $150. Readers can find answers to different riddles by solving subtraction problems on a calculator, turning the calculator upside down, and reading the display. Children will also be intrigued by the number trick that always results in an answer of 198. One final maze involves subtracting multiples of 100 from 10 000 to obtain a goal of zero.

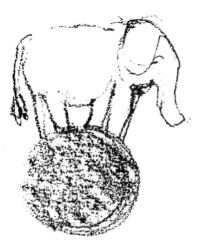

Discovering Shapes

Explorations involving one-, two-, or three-dimensional designs and shapes are the focus of the six-volume set of the Discovering Shapes series. Its format is similar to that of the Discovering Math series, but the selection of the tasks is more concept oriented and should prove more motivating to readers. Each thirty-two-page book is a collection of activities introduced in a one- or two-page format. Each opening activity introduces and defines the topic that is the theme of the book; often real-life connections are depicted or readers are asked to find examples in the world around them. For a number of the activities, materials like paper and pencil, scissors, and tape are needed; often patterns or puzzle pieces must be traced and cut out. Optical illusions, mazes, and tangrams appear in a number of the books, but with a focus or design that highlights the theme of the book. Definitions, additional directions, or other information is often conveyed by "balloon speech" from cartoon characters. Each book contains a table of contents, an answers section, a glossary, and an index.

Riggs, Sandy. *Circles*. Illustrated by Richard Maccabe. New York: Benchmark Books, 1997.

ISBN 0-7614-0458-9 $12.95

☆ Single concept, 3–5

Optical illusions that distort the size of circles and examples of objects in the world around us are included in the opening pages that introduce circles. Drawing a circle is an excellent way to explore its properties, and three methods are suggested in one activity: sketching freehand, drawing around a circular object, or using a cardboard strip for a radius. Other explorations of a circle's properties have students construct circular designs with a compass. In some activities, readers analyze geometric designs to determine which shapes are alike or to determine which design has a greater fractional amount of blue shading. Another task involves determining whether polygons or star polygons are formed as different patterns of points on the circumferences of eight- and twelve-pointed circles are connected. Another fascinating experiment has readers cut apart Möbius strips. Some activities are designed for skills practice but are disguised as games. These activities include a board game to practice identifying chords, radii, and diameters; a counter game to practice identifying concentric and congruent circles; and a counter game to practice using a formula to calculate the circumferences of various sizes of pepperoni on a pizza. In another activity the directions for the "Dot and Dart Boards" are poorly written but become obvious when the answer key is checked. *Circles* contains more practice activities than the other books in this series, but it also contains some good problem-solving activities.

———. *Triangles.* Illustrated by Richard Maccabe. New York: Benchmark Books, 1997.

ISBN 0-7614-0459-7 $12.95

☆ Single concept, 3–5

Colorful quilt patterns and hex signs found on Pennsylvania barns are used to introduce triangles in design; readers are urged to design their own patterns. Another connection to the real world includes examining how the triangular form is used in the construction of bridges; one activity is designed to compare this shape's strength with that of quadrilaterals. *Triangles* also includes toothpick puzzles and origami instructions for folding a penguin from an equilateral triangle. One particularly well-designed task has readers fold sixteen congruent triangles in a given equilateral triangle; the readers are then asked to find squares, trapezoids, and hexagons embedded in this figure. The trianglar pieces from a tangram are used to build other shapes as well as to explore movements in the plane—slides, turns, and flips. Game formats like concentration, tic-tac-toe, and board games have been designed for practicing such skills as recognizing congruent shapes; naming right, obtuse, and acute angles; or drawing scalene, equilateral, and isosceles triangles.

Stienecker, David L. *Patterns.* Illustrated by Richard Maccabe. New York: Benchmark Books, 1997.

ISBN 0-7614-0462-7 $12.95

★★ Single concept, 3–5

Connections to nature, like the patterns in a butterfly's wing or a close-up of a tree's leaf, introduce this topic. Linear patterns are explored through activities that involve constructing a stencil and folding and cutting paper chains; connections are made to the patterns in Native American pottery and weaving. Patterns in two dimensions include weaving with paper strips and exploring tilings that use letters or geometric shapes. Readers are asked to analyze designs for laying bricks, such as the herringbone and double basket weave patterns, or to design their own quilt patterns. Carefully written directions and sequenced diagrams show how to construct intricate patterns inside a hexagon. Other patterns using lines, concentric circles, and curves effectively illustrate optical illusions, mazes, and shimmering designs. One final project has readers draw simple fractal shapes; another asks them to determine whether regular triangles, pentagons, hexagons, and octagons can be used to cover the plane. Stienecker has presented a good collection of activities that are motivating and represent a good cross section of many different types of patterns involving shape.

————. *Polygons.* Illustrated by Richard Maccabe. New York: Benchmark Books, 1997.
ISBN 0-7614-0461-9 $12.95

☆ Single concept, 3–5

From triangle and quadrilateral through ondecagon (eleven-sided) and dodecagon (twelve-sided), a table is used to summarize effectively the names for polygons with different numbers of sides. Regular and irregular polygons are introduced with misleading definitions and diagrams. The diagram to illustrate irregular polygons shows a concave pentagon with unequal angles; additional diagrams are needed to show that irregular polygons can be convex or concave, as well as have equal or unequal sides. Also, the text's descriptions and the glossary's definition are not precise; they state that irregular polygons have sides of different lengths and angles of different sizes, which implies that a rhombus is not an irregular polygon. Other definitions ignore the differences among such concepts as rays, lines, and line segments, for example, "An angle is a shape formed by two lines that meet." A polygon is defined as "a figure with three or more lines joined at the endpoints." Most of the activities are well-designed tasks that are motivating as well as worthwhile. Unique quadrilaterals including squares, parallelograms, trapezoids, rectangles, kites, and rhombuses are defined, then an excellent task that involves connecting the midpoints to form a new quadrilateral is posed. Some of the activities gradually become more complex, such as determining how many different ways two hexagons can be joined, then three hexagons, then four hexagons. Puzzles include tangrams, which readers use to form a parallelogram and a trapezoid. Various shapes of triangles and quadrilaterals are to be traced and cut out for puzzle pieces that can be used to form block letters; guidelines are given for designing block-letter puzzles. Another page shows five different shapes, each of which is to be cut into two pieces to form an explicit shape such as a hexagon or a trapezoid—another good task for developing spatial sense. Analyzing networks is a good task, but the context of traceable polygons is contrived. Other activities include drawing star polygons, playing a concentration-like game, and solving a maze composed of hexagons.

————. *Rectangles.* Illustrated by Richard Maccabe. New York: Benchmark Books, 1997.

ISBN 0-7614-0460-0 $12.95

★★ Single concept, 3–5

Opening pages include a rectangle search and examples of mosaics that are designed to increase our awareness of squares and rectangles in the world around us. Magic squares, strategy games using dot paper, mazes, and toothpick puzzles are some of the interesting activities described; many involve logical reasoning. Another motivating activity has readers fold a container from a square piece of origami paper. Some excellent classic tasks are posed, such as determining the number of squares on a checkerboard or constructing a golden rectangle. Another challenge is to form a square from tangram pieces as well as a rectangle that is not a square. Similar to those in Stienecker's *Polygons,* other puzzles involve cutting a shape and then arranging the two pieces to form a rectangle. Seven pentominoes (consistently spelled *pentaminoes*) are shown with a challenge for the reader to find the other five pieces. In "Rectangle Cards," it is unclear why square playing cards are not used for both activities. Playing cards are used to outline different-shaped rectangles with a constant perimeter; although the perimeter is constant, readers cannot determine the perimeter because the width of the cards is different from their length. When this activity is extended to find different-shaped rectangles with a constant area, square cards are used. Overall, this is a good collection of tasks for exploring shape and developing spatial sense.

————. *Three-Dimensional Shapes.* Illustrated by Richard Maccabe. New York: Benchmark Books, 1997.

ISBN 0-7614-0463-5 $12.95

★★ Single concept, 3–5

An informal but good description of what is meant by three-dimensional shapes begins this book. Pyramids, cubes, prisms, spheres, cones, and cylinders are introduced; each shape's name, with a diagram, appears in a table, and the reader is encouraged to find such shapes in the world around us. Like in Stienecker's *Rectangles,* pentominoes are again used, but this time the task is to predict and then test which can be folded into a box without a top. Later in the book, this task is revisited as readers find patterns that use six squares that can be folded to form a cube. Patterns to be traced and cut out are given for cubes, square pyramids, and octahedrons. Other patterns are for constructing three-dimensional puzzle pieces; a cube can be built from three square pyramids or a tetrahedron from two five-faced solids. Another spatial-visualization task uses four cubes to find all combinations of stacks one, two, three, and four cubes high. (One of the two-cubes-high solutions is missing in the answer key.) The five Platonic solids are introduced through diagrams, descriptions, and tidbits of history; the reader is encouraged to analyze the properties of the Platonic solids and record the results in a table. Other motivating activities involve drawing, diagrams, and illusions. Directions on how to draw three-dimensional shapes are well written and clearly illustrated. Optical illusions include diagrams of three-dimensional objects that are impossible to build or that take on a different perspective as they are studied.

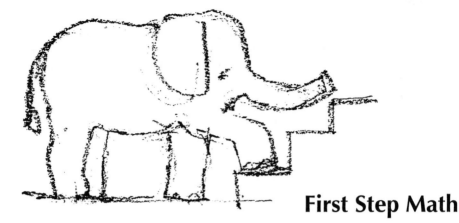

First Step Math

Beautiful photographs clearly complement the simple, direct text of each book in the First Step Math series. Small, sturdy, and attractive, these books will entice the young reader to explore beyond the covers. Opening pages directly connect the mathematical topic to the young reader's world. Each thirty-two-page, colorful book presents and describes interesting activities designed to expand the reader's understanding of the mathematical concepts under study. For the books to be successful, it is essential that an adult interact with the young reader to offer guidance and pose questions as needed. Boys and girls representing a variety of ethnic groups are shown actively involved in the tasks. A particularly strong point of the series is found at the back of each book where there are notes for parents and teachers about the purpose of each task; a page of more things to do; fun facts about each specific topic; a glossary; a listing of places to visit, more books to read, and videotapes to view; and finally, a short index. These additional sections and closing notes extend the learning opportunities. The appeal and high quality of this series are to be commended.

Griffiths, Rose. *Boxes.* Illustrated by Peter Millard. Milwaukee: Gareth Stevens Publishing, 1995.

ISBN 0-8368-1179-8 $13.95

★★ Single concept, 1+ (K–2)

An assortment of containers for cards, dominoes, eggs, and toys is used to motivate children to explore the different shapes and sizes of boxes. Interlocking plastic pieces are used to build boxes that vary from prisms and pyramids to regular polyhedrons. Photographs effectively illustrate size comparisons by showing tiny boxes that contain miniatures and large ones that could be used as playhouses. Size is also effectively shown through an assortment of nested boxes and an experiment that tests how many spoonfuls of popcorn will fill different boxes. Clear and simple directions show how to build triangular and rectangular prisms from cardboard and tape. *Boxes* is an excellent choice for motivating and worthwhile tasks.

————. *Circles*. Illustrated by Peter Millard. Milwaukee: Gareth Stevens Publishing, 1994.

ISBN 0-8368-1109-7 $13.95

★★ Single concept, 1+

Circles opens with the query "How many circles have you seen today?" and color photographs that include objects like clock faces, plates, lids, bracelets, and wheels. Other circular shapes, such as baskets, beads, buttons, and marbles, are appropriately depicted, but the focus of this small book is on circles. Cutting dough with a cookie cutter, dropping a pebble into water, and sitting with a circle of friends are all familiar examples of circles that are part of a child's world. Other clever images include a young girl holding a circular hoop so that it looks like an oval and two children drawing a large circle with a jump rope. This excellent book will help children explore mathematics and build connections.

————. *Facts & Figures*. Illustrated by Peter Millard. Milwaukee: Gareth Stevens Publishing, 1994.

ISBN 0-8368-1110-0 $13.95

★★★ Single concept, 1+

Guinea pigs introduce the topics of measurement and data collection in *Facts & Figures*. Tables, tallies, and graphs are used to collect and record information about voting for the pets' names, scheduling pet care, and deciding what foods the guinea pigs like best. Data on the pets' weights are collected from birth on; this comparison is extended to the reader's weight at birth compared with his or her current weight. Boxes are designed to make suitable hutches; measurement rods are used to design appropriately sized pens. Again, worthwhile tasks are presented through attractive design and clear, simple text. Some "Fun Facts about Facts and Figures" can be shared with the young reader: Guinea pigs are not pigs, but rodents; surveys measure public opinion; and databases can be used to store information.

————. *Games*. Illustrated by Peter Millard. Milwaukee: Gareth Stevens Publishing, 1994.

ISBN 0-8368-1111-9 $13.95

★★ Single concept, 1+

Pick-up sticks, playing cards, and game boards for chess and chutes and ladders are some of the many game-related items shown in the colorful illustrations. Directions for the ancient African game of wari is given, and the appeal of modern computer games is noted. From this brief overview a delightful shift shows children creatively involved in designing their own games; minimum detail is given, but the photographs and the simple text provide enough structure to help the reader get started. Another shift in this thoughtful presentation leads to analyzing games with respect to luck, skill, or both. Questions about another game focus on whether or not both players have an equal chance of winning, a reference to beginning probability. Griffiths and Millard have designed another winning book that presents worthwhile tasks in mathematics to young children.

————. *Number Puzzles.* Illustrated by Peter Millard. Milwaukee: Gareth Stevens Publishing, 1995.

ISBN 0-8368-1180-1 $13.95

★★ Single concept, 1+ (1–2)

Counting backward by twos on the calculator, indicating every third number on a hundred chart, or building each layer of a stair-step pattern with interlocking blocks are some of the number patterns or puzzles presented. In each example, the reader is asked questions about the pattern and must decide what number comes next. A flowerpot used to hide part of a set is cleverly used to introduce part-whole concepts; this format is extended to writing number sentences. Flaps to cover numbers in a sentence and a "function machine" are other creative formats that will motivate the reader to become involved in exploring numbers and addition and subtraction.

————. *Numbers.* Illustrated by Peter Millard. Milwaukee,: Gareth Stevens Publishing, 1994.

ISBN 0-8368-1112-7 $13.95

★★ Single concept, 1+ (K–1)

Number use and numerals are the focus of this beginning book about number. Questions are simply posed and answered about why we have only ten digits. Exploring the shape of different numerals includes such activities as distinguishing between letters and numerals; doing crayon rubbings; and looking at different representations, such as Gujarati numerals, roman numerals, or the numerals from a digital clock. Examples of how we *use* number are also presented; colorful photographs show children using scales and rulers to measure weight and length. House numbers, telephone numbers, and numbers to indicate clothing sizes are also effectively illustrated. Griffith and Millard's presentation is effective in connecting the many representations of numerals and number use to a child's world.

————. *Printing.* Illustrated by Peter Millard. Milwaukee, Wis.: Gareth Stevens Publishing, 1995.

ISBN 0-8368-1181-X $13.95

★★ Single concept, 1+ (K–2)

   Books, fabrics, table napkins, and mailed letters all have one thing in common—they have been printed either by machine or by hand. With these connections to our world the reader is introduced to a number of activities that involve making prints by using stamps or common items. Different patterns can be formed from the same print block by geometric movements referred to as translations or rotations. Reflections are also explored as readers decide which side of a print block to paint so the stamp will print the letters correctly. Very young children may have difficulty with the reverse print. These simple, but well-designed tasks will motivate the young child to begin explorations in geometry.

————. *Railroads.* Illustrated by Peter Millard. Milwaukee: Gareth Stevens Publishing, 1995.

ISBN 0-8368-1182-8 $13.95

★★ Single concept, 1+ (K–1)

   Straight lines, curved lines, and simple and closed curves are some of the geometric concepts explored through the design of toy railroad tracks. Dilemmas are illustrated; for example, could a train travel along a track that contains a right angle? The children in the photographs are shown building with Legos, blocks, and paper. Apartments, factories, and other buildings are constructed and connected by railroad tracks; bridges and tunnels are carefully designed so that the train can pass under them. In addition to building a model railroad and connecting communities along its route, the children are also shown drawing maps that reflect the layout of their three-dimensional community. The notes to the parents or teachers are most helpful in connecting the activities to mathematics.

I Love Math

I Love Math is an appropriate name for this exciting series! Designed to place mathematics in meaningful and challenging contexts, the series adopts a playful tone in presenting humorous short stories, riddles, and games that have mathematics concepts and problems embedded within them. In each book, an introductory letter to parents includes a general explanation of mathematics, how childen learn it best, and what parents and their children can expect to find in the book. The inclusion of parents in this learning partnership is reinforced through the comments for parents that are found at the bottom of many pages. Suggested questions that parents might pose, materials needed for participation, and possible extension activities are provided. The table of contents lists each provocative story, game, or riddle title and includes a brief description of the mathematics processes and concepts developed within each. All the books are multiconcept, and they have been liberally cross-referenced throughout *The Wonderful World of Mathematics*. A different theme is explored in each sixty-four-page book: the zoo, space, the sea, amusement parks, the city, food, the body, the mall, sports, the house, and make-believe. The interactive, hands-on problems posed around each theme highlight the presence of mathematics all around us.

Consultants to the series include mathematics specialists, experienced teachers, curriculum specialists, and theme specialists where appropriate. They have carefully designed the series to revisit certain mathematics concepts with greater sophistication to acknowledge the growing understanding of their young readers. In each book at least one activity requires a calculator. Riddles involving logical reasoning are often included, and problems are presented through poetry and stories. As children work their way through the series, they will begin to look for the special mystery story that features Professor Guesser, Problem Solver, whose detective work becomes increasingly difficult and complex as the series progresses.

Appropriate mathematics vocabulary is introduced and reinforced throughout the series. Inside covers of each book contain simple, but theme-specific games that call for such everyday materials as different-colored buttons or doughnut-shaped and square-shaped cereal pieces. The artwork and layout are attractive, colorful, and eye-catching. A variety of artistic styles are represented; all are appropriate. The regular binding is not strong and may not withstand the frequent use that this motivating and attractive series may expect to see. The variety of expertise represented by the editors and staff of Time-Life for Children and their consultants has successfully produced a series that will excite, challenge, and motivate parents and children alike.

Time-Life for Children Staff. *Alice in Numberland: Fantasy Math.* Alexandria, Va.: Time-Life for Children, 1993.

ISBN 0-8094-9978-9 $16.95
ISBN 0-8094-9979-7   (library binding)

★★★ Multiconcept, 1–4 (K–3)

Alice's adventures in Numberland begin when she falls into a tulip bed and meets Number Card Four, a playing card, dressed in a jacket with four pockets and four buttons and emblazoned with the numeral 4. This guide introduces Alice to several riddles all involving the answer 4: the four seasons, addition and subtraction facts, dominoes, the fourth month and four quarters, and a silly riddle on whether a four-foot petunia or a four-foot pickle is taller. Later in the book similar measurement and number activities center on seven, twelve, and sixteen. Estimation and capacity are the mathematical focus of a story involving King Inchandfoot, Queen Yardly, Prince Fillup, and Prince MT as they contemplate how to fill a large tub. Baking a cake in the palace kitchen is the setting for an introduction to fractions and division. The cake decorated with twelve roses is cut into halves, thirds, fourths, and sixths as additional guests arrive. Each time, a diagram clearly shows the fractional pieces and poses questions about the number of roses on each piece. A sequential activity involves folding paper circles into fractional parts. Building houses from cubes for Block Boulevard, identifying shapes in a stained-glass window, and exploring symmetry with a mirror are some of the geometry concepts visited. Eight pages focus on the number ten: riddles about the number ten, representations of different numbers with strings of beads to show grouping by ten, and place-value patterns and puzzles involving the hundred chart. All are thoughtfully designed. Alice's visit to Symbolville is used to introduce the concepts and symbols for comparison: equals, less than, and greater than. Readers help Professor Guesser with her mystery story by using a calculator to solve a long addition and subtraction sentence containing large numbers.

———. *The Case of the Missing Zebra Stripes: Zoo Math*. Alexandria, Va.: Time-Life for Children, 1992.

ISBN 0-8094-9954-1 $16.95
ISBN 0-8094-9955-X   (library binding)

★★★ Multiconcept, 1–4 (K–3)

Measurement is the focus on this visit to the zoo; with each measurement challenge, readers are encouraged to develop and use their powers of estimation. For example, estimating how many pink flamingoes on a two-page spread is the first zoo stop; suggested ranges for the reader to consider are 10 to 20, 100 to 200, or 1000 to 2000. To help estimate, the parent notes suggest first estimating how many pink flamingoes could be covered by one hand. To measure a snake's curvy length, string is needed, which can then be compared with the twelve-inch ruler on the base of the page. This format is repeated with various animals; each time the reader is encouraged to estimate first. Professor Guesser collects data at different time intervals to solve "The Case of the Missing Zebra Stripes." Time estimates and concepts of a minute are involved in such activities as "Can you stand on your leg for one minute?" Logical reasoning is needed to solve a series of riddles about the zoo's newest arrival; each illustration shows a child thinking through each clue. A bar graph organizing the data for each giraffe's height is used to solve a problem of feeding trays. Straight, curved, parallel, and intersecting lines are the basis of explorations for design, circles, and polygons. Operations and place-value concepts are embedded in solving the problems in "Henry's Tale." A calculator is a helpful tool to solve the alphabet code values for the names of different zoo animals, but its use appears forced for "How does an alligator use a calculator?" Adding 5s, 10s, and 3s is involved in one of the board games on the book's end pages; recognizing odds and evens is embedded in the other.

———. *From Head to Toe: Body Math*. Alexandria, Va.: Time-Life for Children, 1993.

ISBN 0-8094-9966-5 $16.95
ISBN 0-8094-9967-3   (library binding)

★★★ Multiconcept, 1–4 (K–3)

A variety of measurement and number concepts are developed through body measures in this motivating volume. "Are You Square?" is the aptly chosen title for one of several investigations into the measures of, and facts about, the human body. After measuring and comparing his or her height and arm span, the reader is encouraged to measure different family members to determine who is a square and who is a rectangle. Another activity, "Measure for Measure," is written in script form for the reader and other family members to act out as characters in a play. The focus of these opening activities is to engage the reader actively in measuring and comparing different lengths, such as head circumference, waist, spine, and leg. Strips of cardboard are used to compare the measures, not units of length. Eventually all these "measures" are used to construct a skeleton model of the reader. Reading a scale denoting inches and feet is the focus of a story about Kate's yearly visit to an amusement park; the reader is to determine how many inches Kate needs to grow each year until she is four feet tall—the height requirement for the snake ride. In the "Case of the Crooked Bookcase," Professor Guesser solves the problem by examining nonstandard measures and the need for standard measures, which leads to estimation activities that use a foot's length. Grouping and combinations are explored with fingers and counters on five-frames and with a game on ten-frames. How to *sign* numbers from 1 through 20 is cleverly presented and explained. Operations with numbers are involved in answering questions about different groups of bones in the human body and facts about the digestive system. Senses are the basis of some activities like "Eye See" (optical illusions involving spatial sense and estimation) and "Tongue Twister" (problems with measures). A pictograph presents data on how many cups of blood a body has compared with its weight; the reader is asked to analyze the data. The calculator is a welcome tool for determing one's age in days or hours.

————. *The House That Math Built: House Math.* Alexandria, Va.: Time-Life for Children, 1993.

ISBN 0-8094-9986-X $16.95

★★ Multiconcept, 1–4 (K–3)

This member of the I Love Math series will really appeal to the primary set! Its humorous approach includes the silliness of a two-act play involving a function machine; the reader is invited to predict what will come out of a food-vending machine, the Kitchen Magician 5000, given various pieces of data. Logical thinking is developed as readers use clues to match several dogs to the appropriate dog houses. The calculator becomes a tool for solving number tricks involving ZIP codes. Colorful, stylized illustrations take the reader through a series of rooms in "the house that math built." The journey through the rooms develops specific mathematics vocabulary for length, shape, time, patterns, addition, and subtraction. A mystery format is used to develop logical reasoning, patterning skills, and spatial sense in both "Who Lives in This House?" and "2B or Not 2B?" A variety of data are embedded in these stories as the reader is asked to determine the identity of the residents in the first mystery and the exact location of an apartment in the second. Quilting is introduced as a context for plane shapes and patterns as a grandmother describes the process of piecing a quilt. Readers are encouraged to develop their own quilt pattern by using variations of the nine-patch design. Similarly, tiling and arrays using squares are involved in an excellent problem-solving story starring Professor Guesser; she helps her neighbors settle the dispute over who has been assigned the garden plot with the greatest area. Packed with unique problems in real-life contexts, this volume will motivate readers to generate additional problems around their own house and neighborhood. *The House That Math Built* connects logically with the *Look Both Ways: City Math* volume of the series.

————. *How Do Octopi Eat Pizza Pie: Pizza Math*. Alexandria, Va.: Time-Life for Children, 1993.
ISBN 0-8094-9950-9 $16.95
ISBN 0-8094-9951-7   (library binding)
★★ Multiconcept, 1–4 (K–2)

Fantasy, food, and animals abound in each delightful tale. Octopi eating pizza pie is the story line as each piece is removed from the pie, a circle divided into eight equal pieces. Follow-up questions reflect both what is left as well as how many pieces each octopus is holding. Later in the book, circular and rectangular shapes are used to explore halves and fourths. In "Elephant Juice," poetry describes the adventures of Hector and Ellen as the two elephants compare capacity measures. Other capacity activities include comparing different-shaped containers and estimating the number of goldfish crackers in a bowl. A function machine is cleverly introduced and serves as the basis for several activities; the mathematical functions include adding, doubling, and subtracting. In Professor Guesser's "Hold the Anchovies," geometric figures mysteriously change from two- to three-dimensional shapes when baked by Pepper Pepperoni, pizza proprietor. Tallies in groups of five are introduced in "How Do Butterflies Count French Fries?" Logical reasoning, addition, and subtraction are needed to solve various riddles as well as a story about a scarecrow and his garden. Simple addition and subtraction are involved in calculator magic tricks in which the answer is always the beginning number. Readers are encouraged to determine why the tricks work, try them with other family members, and create their own tricks. Both game boards at the end pages involve addition and ... pizza!

————. *Look Both Ways: City Math*. Alexandria, Va.: Time-Life for Children, 1992.
ISBN 0-8094-9958-4 $16.95
ISBN 0-8094-9959-2   (library binding)
★★★ Multiconcept, 1–4

Readers explore a city by using a variety of mathematical skills and concepts. Mathematical ideas presented in interesting contexts include examining a variety of shapes and patterns found in the structure of a city, using coordinates to find humorous events in an apartment building, and using number patterns to complete missing building addresses. Ordinal numbers are embedded in a series of logical problems. A fairly sophisticated mystery involves readers in helping the Math Patrol use map skills to find a hidden treasure. Another challenging problem asks the reader to keep track of the number of people getting on and off a city bus along its route (manipulatives are encouraged!). The geometric shapes and patterns found in the city scenes and structures are captured both in photographs and in fantasy illustrations; three-dimensional shapes including spheres, pyramids, and rectangular prisms are represented and named. Once again, calculators are used to solve various problems in "How Does Peter Feed the Meter?" To compare distances of different travel routes, children are encouraged to use a simple road map and a calculator; one example of key strokes is given. Throughout *Look Both Ways: City Math,* realistic and interesting problem situations are presented.

———. *The Mystery of the Sunken Treasure: Sea Math.* Alexandria, Va.: Time-Life for Children, 1993.

ISBN 0-8094-9994-0 $16.95

★★★ Multiconcept, 1–4 (K–3)

From sea creatures to sunken treasures, the sea is the setting for a number of excellent mathematical tasks. In "Fishy Party," photographs of six colorful fish are the object of questions relating to fractions as the reader considers different parts of the group. The sandy seashore is the location for building sand castles of various shapes and arranging seashells to represent number sequences. One pattern is the square numbers and the second one is the triangular numbers. In each instance the reader is asked to predict the next number in the sequence. Escher-like drawings of fish and seahorses introduce the art of tessellations; clear diagrams and directions are given for making tessellations by using translations. A colorful illustration of creatures of the sea floor leads to questions involving data; the number of each type of creature is displayed in a horizontal pictograph as well as in a vertical bar graph. Professor Guesser teams up with Jacklean Mouseteau to solve the "Mystery of the Sunken Treasure." The solution to the problem is found by doubling pieces of gold; an illustration dramatically shows the one, two, four, eight, sixteen, ... pieces for each of fourteen days. A photograph of a blue lobster is the subject for estimating measures of length. Tangrams are introduced in a story form similar to *Grandfather Tang's Story;* a pattern for the tangram puzzle is included. Pictographs on which each symbol can represent a value greater than 1 are introduced in "Graph Their Muffins," a story told in poetry form. Calculators are used for finding neighboring numbers that have a specific sum, for example, $14 = 2 + 3 + 4 + 5$. Throughout this book the activities are very thoughtfully selected and designed.

———. *Play Ball: Sports Math.* Alexandria, Va.: Time-Life for Children, 1993.

ISBN 0-8094-9970-3 $16.95
ISBN 0-8094-9971-1   (library binding)

★★ Multiconcept, 1–4 (K–2)

Measurement, statistics, and probability are some of the topics encountered in the mathematics of sports. An opening story told in verse has Ellen and Hector, the elephant twins, at a ball game. Operations and fractions are encountered as they purchase and share food; data are analyzed as they make sense of the score chart. The black-and-white design of a soccer ball leads to an activity to determine which color a player would more likely touch at random; the geometric designs of pentagons and hexagons are also considered. Probability is the focus of deciding who starts the game with activities like tossing the bat or flipping a coin. Reading scales is accomplished by reading the markers at the swimming meet and considering the depths at different parts of a swimming pool. Photographs of individual children are featured beside a data card that names the sport in which he or she is involved and gives other relevant data; the reader is asked various questions about each situation. In "A Cheetah in the Fast Lane," Professor Guesser solves the dilemma of how to make a race fair when racers are running in four lanes around an oval track; the clever story and the well-designed illustrations clearly show the process of measurement. "Bull's Eye" involves adding different numbers to obtain a specific score. Logical reasoning and Venn diagrams are used for sorting different objects. Interpreting graphs is involved in "Amazing Animal Athletes"; speeds of different animals in miles per hour are given in a bar graph marked in units of ten. Further comparisons are made between animals and people for events like high jumps and swimming speeds. An appropriately active closing activity is "You Try It," featuring six events involving time, length, and estimation. One event focuses on the amount of time needed to jump over seven hurdles—empty milk containers.

———. *Pterodactyl Tunnel: Amusement Park Math.* Alexandria, Va.: Time-Life for Children, 1993.

ISBN 0-8094-9990-8 $16.95

★★★ Multiconcept, 1–4

Pterodactyl Tunnel is a natural extension to using games to place mathematics in an enjoyable context. In this book, readers find themselves involved in more than games; they are right in the middle of an amusement park with its variety of enjoyable opportunities. Trying to locate someone in an amusement park can involve both number and logic. These concepts are needed to solve the problem of "Thunder Park" and to help Professor Guesser find the missing owner of a raffle ticket. As readers move on to the rides, they use multiplication and addition skills to determine how many can ride on a turn or how long before they have their own turn in the minute mystery, "Showtime." Patterns help readers navigate the waters of this volume's theme ride, "Pterodactyl Park," and determine the winning animals for the "Great American Racing Derby." What amusement park participant could resist the arcade? It's loaded with games from brick dance (a little like Tetris) to sums on a triangle's edge, Pascal's triangle, and traceable networks. Determining different combinations comes into play as participants design masks at the "Make-a-Mask" booth. Calculator use is encouraged at the balance booth; this problem-solving page will be enhanced if the reader actually uses a balance to verify the number sentences solved on the calculator. Opportunities in this volume include many short problems that develop a wide variety of skills and concepts in a format that is fast-paced—just what you expect from an amusement park!

————. *Right in Your Own Backyard: Nature Math*. Alexandria, Va.: Time-Life for Children, 1993.

ISBN 0-8094-9962-2 $16.95
ISBN 0-8094-9963-0   (library binding)

★★ Multiconcept, 1–4 (K–3)

Plants and animals are the vehicle for the numerous mathematics topics encountered in *Right in Your Own Backyard*. Illustrations of flowers from countable to a large field provide an opportunity for beginning estimation strategies. A pastoral scene is the basis of a logical reasoning activity for finding animals with specific characteristics. Geometric shapes of spirals, concentric circles, stars, hexagons, and branching patterns are introduced on attractive two-page spreads that include diagrams, a poetic format, and photographs of these patterns as they are found in nature. Shapes including polygons, star polygons, and other designs are explored in an activity that involves connecting points. Another geometric activity uses a mirror to explore the symmetric structure of various plants and animals. Professor Guesser uses patterns, statistics, the calendar, and logical reasoning to solve "The Trash Can Caper." Temperature measurement is the mathematical focus of a story about ants and their year. Data about rainy days are collected on a yearly calendar and displayed on a picture-bar graph; questions about the data are included, and readers are encouraged to devise further questions and collect their own data. More data about the lengths, weights, and heights of animals are the basis for estimation questions for developing number sense. Beautiful close-ups of flowers are the basis for addition problems. Skip counting or repeated additions are facilitated with a calculator's constant function. The attractive photography enhances the reality aspect of *Right in Your Own Backyard,* and the appealing cartoonlike illustrations convey the needed information, questions, or story line.

————. *The Search for the Mystery Planet: Space Math.* Alexandria, Va.: Time-Life for Children, 1993.

ISBN 0-8094-9982-7 $16.95
ISBN 0-8094-9983-5   (library binding)

★★ Multiconcept, 1–4

Set in a context that is a little further removed from the readers' concrete personal experiences, this volume's space theme nevertheless has great interest for young mathematicians! Connections with the scientific community are another plus for the context of these problems, which range from highly factual to science fiction. Data analysis is the key focus of the first series of problems, which center on facts about the planets in our solar system. Readers are encouraged to connect their observational skills with their understanding of the calendar and elapsed time as they chart the phases of the moon in a problem that will take a month to complete fully. Patterns and shape are the focus of two activities that invite readers to identify familiar constellations in the night sky and to create a personal star pattern. A good activity for examining patterns of change and developing spatial sense is "My Shadow," in which body tracings at various times of the day show the different lengths of shadows as the position of the earth in relationship to the sun changes. A series of blips and beeps keeps Professor Guesser involved as she cracks a code that allows the astronauts on a space mission to decode correctly a message warning them of a space-related hazard. Follow-up problems for codes are given at the end of the story. As with many problem-solving situations using a space-related theme, this volume is forced to translate highly abstract concepts into user-friendly contexts. Although the format and the problems are motivating and entertaining, the problems more often develop skills with numbers and calculation than the space concepts. For example, calculating the differences in a person's weight on Earth and on the Moon assists children in realizing that their weight would be different on the Moon, but it does not develop an understanding of the underlying concepts. Likewise, a discussion about Neptune's variant orbit describes it (perhaps necessarily) as "going around on its side" instead of explaining the different orientation of its poles. In spite of the inherent drawbacks of working with astronomy concepts with young learners, this book is motivating and contains many fine opportunities to integrate mathematics with science.

————. *See You Later Escalator: Mall Math.* Alexandria, Va.: Time-Life for Children, 1993.

ISBN 0-8094-9974-6 $16.95
ISBN 0-8094-9975-4   (library binding)

★★★ Multiconcept, 1–4 (1–3)

See You Later Escalator introduces a variety of mathematics topics to show that much more than shopping takes place at the mall!

"Midnight at the Mall" is a story that combines fantasy and reality with after-hours activities; the identification and functions of right, obtuse, and acute angles are highlighted in the mall's architecture and in different objects, such as folding chairs and stepladders. Professor Guesser uses spatial sense and charts to draw a directory of the mall after it is determined that no one can find the lost and found department. Spatial sense is also involved in a crowded scene where the reader is asked to give directions to specific locations. A chart for determining shoe size is introduced in "Sneakers." Mall walkers provide the context for questions on estimation and time. Combinations are cleverly presented in "Charlie the Chooser" as Charlie and Maria contemplate what choices can be made; the illustrations clearly show tree diagrams as a format for solutions. "The Food Court" and "The Candy Corral" provide situations dealing with operations and money. At the arcade, questions are posed to determine which numbers can be combined to earn a specific score on a bull's-eye; this is extended to different combinations for 100. "More than one question for each answer" and number use are highlighted in a mathematics contest in which West Decimal Elementary is pitted against Computation Country Day. One game has readers add 10 to a single-digit number on a spinner; adding multiples of 10 is involved in a game whose board is included at the back of the book.

Let's Investigate

ISBN 1-85435-773-5 (Complete Set)
ISBN 1-85435-455-8 (Group 1) $101.70
ISBN 1-85435-463-9 (Group 2)

Fifteen to thirty investigations focus on a specific topic in each sixty-four-page book in this series. In each of the books' sections various questions are posed. Most of them initiate problem solving; some are skill oriented. A number of excellent classic problems are incorporated in the investigations. Puzzles and games are used to address both concepts and skills; although many are very well done, others seem contrived. Connections to history and real-life situations are incorporated throughout each book. The text moves smoothly and thoughtfully from one section to another. Color contributes to the clarity and appeal of the effective diagrams and drawings. Some of the volumes in the series are written so that a reader who makes an error can work with additional examples; otherwise, the reader is encouraged to continue with a different page. Answers are listed in the back of each book or a reference is given to specific pages that completely describe the solution process. The investigations are interesting and worthwhile, but the format of the books is more textbooklike compared with children's books like *Math for Smarty Pants.* Common features for each of the eighteen books are a table of contents, a glossary, and an index. In the later additions to this series, the opening notes indicate what materials are needed to work through the investigations. Group 1 includes the titles *Circles, Numbers, Number Patterns, Quadrilaterals, Area and Volume,* and *Triangles.*

Smoothey, Marion. *Angles.* Illustrated by Ted Evans. New York: Marshall Cavendish, 1993.

ISBN 1-85435-466-3 $17.95

☆ Single concept, 4–8 (5–8)

The opening of a hinged-box's lid, the movements of a clock's hands, and the rotations of a merry-go-round or a key in a lock— these turns and fractional parts of turns are some of the connections used to introduce the concept of angles. Another application involves semaphore, a language constructed from the angles formed by a user's arms and body to send messages. Quarter, half, and three-quarter turns are effectively connected to measures of 90, 180, and 270 degrees by an activity involving folding circles into fourths. This model is extended to eighths and is also related to the twelve divisions of a clock's face. Two thin strips of cardboard and a paper fastener are suggested for building a model of an angle, but language in the text is imprecise because an angle is described as formed from two lines rather than from two rays. Estimating angle size is appropriately introduced prior to exercises for measuring angles with a protractor. Special angles, such as straight, acute, right, obtuse, and reflex, are defined and are included in classification activities. Constructions for drawing perpendicular lines and bisecting angles with compass and straightedge are given but are not incorporated into other activities. An activity on constructing similar triangles by drawing parallels is appropriately followed by an application that uses a clinometer to measure heights. Another application includes making a periscope by considering light reflection. If additional practice is needed for some tasks, such as estimating angle size, the text is written so that the reader has the options of practicing more or continuing to new activities. Overall, some good connections and explanations are included.

―――. *Area and Volume.* Illustrated by Ted Evans. New York: Marshall Cavendish, 1992.

ISBN 1-85435-460-4 $17.95

★★ Single concept, 4–8 (3–8)

Comparing areas of seven different figures is the opening activity for *Area and Volume.* For checking the estimate, the next page shows the same figures, but they have been subdivided into congruent triangles so the areas can be counted with ease. Other arbitrary units, such as rhombuses and hexagons, are used to cover different figures before the square is introduced as a common unit of measure and before estimating and counting parts of a square unit are described. Area formulas for rectangles are connected to counting by considering the number of square units in one row and multiplying by the number of rows; this procedure is then generalized to length times width. Formulas for areas of parallelograms and triangles are developed through paper-cutting activities. Separate sections include problems with fractional linear units and lengths of different units. Puzzles involving spatial perception include rearranging parts of shapes to show that the area is constant even though the figures look different. Another investigation uses different lengths of drinking straws to show that constant perimeter can result in different areas;

this is extended to maximizing area. Connections to the uses of area in our daily lives include determining how much grass seed should be spread on the lawn and the cost of a slab for a patio. To introduce filling space or volume, a grocery display of stacked cans of beans is effectively used. Front views of the display show how many cans are in each row and how many rows high the display is; a side view shows how many rows or layers deep it is. Counting cubes and considering layers are extended to finding the volume of prisms. Connections are made between the ratio of a cube's surface area to its volume and the types of animals that can exist in cold climates. A final investigation explores how volume changes when the linear dimensions double; an earlier investigation similarly explores the linear and area relationship.

———. *Calculators.* Illustrated by Ann Baum. New York: Marshall Cavendish, 1995.
ISBN 1-85435-777-8 $17.95
☆ Single concept, 4–8 (3–8)

Fingers and toes, abaci, slide rules, and Napier's rods are cited as early calculating tools with brief notes about their historical uses. Directions for making Napier's bones and simple slide rules are carefully illustrated and explained for the purpose of understanding how these tools were used. The remaining forty pages deal with electronic calculators. Distinct sections explain how to use the clear key, memory keys, and constant function and, for added interest, include familiar games such as 21 and calculator word riddles. Readers are challenged to express different values by using only one digit several times with various operations; another challenge involves the fact that 43×68 and 34×86 have the same product and asks the reader to find other two-digit factors that have the same product when the digits are reversed. Some activities explore the calculator's display from beginning ideas like entering $14.40 to more sophisticated ideas like determining what happens with extremely large or small numbers. Rounding numbers, estimating answers, finding remainders, and working with fractions and decimals are included. When discussing how to key codes, Smoothey addresses the order of operations and mixed numbers, but later, to solve 8 4/7 − 3 2/3, the text reads "Key in 8 + 4 ÷ 7 − 3 − 2 ÷ 3 =." Some errors in the text, like missing operation signs or wrong numbers, are inconvenient, but not problematic. *Calculators* contains activities and information that vary from fair to excellent.

———. *Circles.* Illustrated by Ted Evans. New York: Marshall Cavendish, 1993.
ISBN 1-85435-456-6 $17.95
★★ Single concept, 4–8 (3–8)

In introducing the parts of a circle, Smoothey relates circumference to *circum,* the Latin word for *around,* and radius to *radiate.* Initial activities include folding a circle to find its center. Techniques involving a thumbtack, cardboard strip, and a pencil; a string and a pencil; and a compass are used to draw circles and explore a circle's attributes. To replicate the attractive designs and patterns shown throughout the book, the reader will be involved in creative explorations of the circle's properties. Paper folding is used to explore different shapes that can be derived from a circle; these shapes include an ellipse and a square. A compass is used to create a cardioid, a nephroid, and cusps; patterns of lines are also used to create an envelope of these shapes. Another construction includes determining which regular polygons or star polygons can be formed from eight- or twelve-point circles. Measuring activities lead to the generalization that a circumference is slightly longer than three times the diameter. Connections are made to early uses of this relationship and to the different values of pi over time. Counting squares in a circle is used to approximate the area of a circle, and paper-cutting activities are used to derive the formula. Final pages describe the importance of circles in our culture and in nature.

———. *Codes and Sequences.* Illustrated by Ann Baum. New York: Marshall Cavendish, 1995.
ISBN 1-85435-774-3 $17.95
★★ Single concept, 4–8 (3–8)

Julius Caesar's simple substitution codes effectively introduce the uses of, and the patterns for, creating codes in *Codes and Sequences.* Directions for constructing a cipher wheel are given so readers can decipher some of the beginning codes or create their own. Some activities include random codes, code words, and nulls or use numbers and number grids. To help decipher codes, a bar graph of letter frequency for the English language is given. Activities are designed that explore these averages and are extended to letter frequencies for French and German. The introduction to factor grids is a nice extension for the study of factors. An investigation on how Braille is coded leads to determining the number of ways that dots can be placed on a three-by-three grid. A similar investigation explores combinations of dots and dashes in Morse code. The final pages mention such connections as bar codes and patterns in music. Investigations in the book provide a good balance of problem solving, connections, and extensions in mathematics.

———. *Estimating.* Illustrated by Ann Baum. New York: Marshall Cavendish, 1995.
ISBN 1-85435-779-4 $17.95
★★ Single concept, 4–8 (3–8)

Differences between exact numbers and estimates introduce

applications and strategies for estimating. Recognizing estimates is the focus of ten diverse examples to broaden readers' view of estimates, which are defined as intelligent guesses. Strategies for obtaining good estimates are embedded in various investigations. To estimate the number of people in a stadium, sections are suggested; for the number of candies on a plate, groups are helpful. For estimating the number of books in a bookcase, the number of books on one shelf is considered. Each investigation is carefully structured and well illustrated to help the reader devise efficient strategies and, consequently, good estimates. Questions about the number of leaves on a tree or hairs on a head introduce an investigation on estimating large numbers. Strategies like rounding and pairing nice numbers are introduced through a conversation between two individuals. Other strategies embedded in the investigations include grouping similar numbers, using place value, and looking at patterns. Some situations involve estimates for adding numbers; others include multiplying by 10s and 100s. The last fifteen pages involve estimating measures. For one series of illustrations, a person is shown next to a small tree, a large tree, a redwood, and then a newly planted bush; in each example, comparison is used to estimate the heights of the trees. These strategies are extended to volume as well. *Estimating* is a good overview of estimation strategies and the ways that estimation is used in our lives.

———. *Graphs.* Illustrated by Ann Baum. New York: Marshall Cavendish, 1995.

ISBN 1-85435-775-1 $17.95

★★ Single concept, 4–8

Grid paper and dot paper are essential for the many graphing activities investigated throughout the book. Conversion graphs to change miles to kilometers are the first graphs introduced; the main function of this activity is to read information from a line graph. The next set of investigations should intrigue the reader and are excellent ways to understand the horizontal and vertical axes; the task is to match a line graph to stories like Rosa's journey to school or the water level in a bathtub. For each story, readers are given choices of potential graphs and they are to explain why some graphs will not work. To introduce curved graphs, the path of a thrown ball is traced. The importance of scale, labels, and titles is discussed; plotting points to determine both line and curved graphs is also introduced. Plots of height versus waist measure is the context for introducing scatter graphs. A second scatter graph represents the height of different children versus the number of their pets to illustrate that it would not make sense to discuss this relationship by drawing a line of best fit. A game introduces coordinate pairs. Another game is used to introduce equations of lines by plotting a sum of 6 when a red die (x-value) and a green die (y-value) are tossed. The coordinates of the various sums are connected even though only integer values can be generated by the dice. Ordered pairs are extended to include both negative and positive values as well as fractional values. As noted above, a variety of good topics and investigations are presented in *Graphs*. Reproducible copies of grid and dot paper are contained in the book.

————. *Maps and Scale Drawings*. Illustrated by Ann Baum. New York: Marshall Cavendish, 1995.

ISBN 1-85435-778-6 $17.95

★★ Single concept, 4–8 (3–8)

Mazes, maps, and other scale drawings are the context for orientation and similarity applications. An opening page introduces the complexities of giving directions from one location on a city scene to another by considering different orientations. In one investigation the text's directions do not match the diagram; the reader is to give directions to Ethan to get through the maze, but the diagram does not include Ethan's location. Initially directions are restricted to right and left or straight ahead; later investigations include using compass points as well as making a compass. Exploring the use of map legends includes asking readers to match maps to scenes of a city. Reading locations from grids is extended to six-figure grid references, which require a knowledge of decimal concepts and estimation. Reading a map's scale is extended to the problems of projection in maps of the earth. Excellent examples of a spherical surface include the difficulties of laying an orange peel flat, illustrations of Goode's projection, and graphing the outline of a dog on a stereographic projection and then on regular grid paper. Other investigations of scale drawings include designing floor and yard plans, reading a plan for building a letter rack, and analyzing models of three-dimensional buildings. The use of survey instruments to measure angles is also connected to scale drawings and determining distances. Despite some minor errors and a misguided example of explaining "right" versus "left" by noting that "right" is the hand "you write with," *Maps and Scale Drawings* contains a number of worthwhile mathematical tasks and interesting connections to history.

————. *Number Patterns*. Illustrated by Ted Evans. New York: Marshall Cavendish, 1993.

ISBN 1-85435-458-2 $17.95

★★ Single concept, 4–8 (3–8)

Rectangular, square, and triangular numbers, as well as patterns like Fibonacci's sequence and Pascal's triangle, are well developed in this book. Arrays are used to explore different rectangular numbers and their relationship to square numbers. Investigations clearly show how to relate square numbers to the sum of odd numbers; another investigation explores the relationship between triangular and square numbers. The "Number Cruncher," a function machine, explores different rules with respect to input and output. Both three-by-three and four-by-four magic squares are included in problem sets and traced to their historical origins. Examples from science involving very small and very large numbers are used to introduce writing numbers with scientific notation; this study of exponents leads to patterns from the classic story of the number of grains of rice on a chessboard. A number of patterns are investigated that involve the Fibonacci sequence. Patterns with coin tossing and maze routes lead to explorations of Pascal's triangle. Problems posed with various patterns on a hundred grid are interesting, but one chart on page 48 is not colored correctly and consequently cannot be solved. New problems called windows, formed from part of a hundred chart, are intriguing, but using the game hangman to record each incorrect guess is a questionable format. Another familiar chart, the calendar, is the focus of a final exploration. Some of the patterns lead to algebraic generalizations, such as moving eight counters in a row. In spite of some confusion from format errors, *Number Patterns* is a good collection of classic problems.

————. *Numbers.* Illustrated by Ted Evans. New York: Marshall Cavendish, 1993.

ISBN 1-85435-457-4 $17.95

☆ Single concept, 4–8

Symbols from the ancient Mayan, Greek, Roman, Babylonian, and Egyptian numeration systems are used to explore the concepts involved in creating numeration systems. Included are other number bases, place value, and the importance of zero. These discussions are extended to adding and subtracting numbers on different types of abaci and the use of the binary system in computers. Regrouping is clearly shown on one abacus, but the modern Chinese and Japanese abacus is simply used to represent numbers after mental calculation. These activities, as well as those for adding binary numbers, may be difficult to follow if the reader does not have some prior knowledge of bases, since descriptions are brief and somewhat procedural. Other types of numbers discussed in the last two-thirds of *Numbers* include odds and evens, primes, fractions, decimal fractions, and negative numbers. Odds and evens are well defined by matching pairs of counters, and well-placed questions lead to generalizations for adding odds and evens. The sieve of Eratosthenes is included in the investigations on primes; multiples and factors are included in the discussions but composites are not mentioned. The presentation of primes is uneven: on page 30, it is stated that 27 is divisible by 3 *because* the sum of the digits is 9. On the next page a mathematical argument is given for why no largest prime exists. Many interesting diagrams are used to explore dividing wholes into fractional pieces and finding equivalent fractions; the puzzlelike investigations are good but are limited to congruent pieces. Finding equivalent fractions and ordering fractions are also presented by plotting ordered pairs on a coordinate grid. A strangely labeled number line is used to introduce decimal fractions; 3.176 is inappropriately read as "three point one seven six." The four pages that constitute this section are not well developed. Real-world problems are used to introduce negative numbers and addition of integers. A number line and a board game are included in two of the activities. The reader is encouraged to use patterns to fill in a subtraction table as well as a multiplication table. Largely because of the broad range of topics, the treatment in this book is inconsistent with respect to the development of the content and the expectations of the readers' backgrounds.

————. *Quadrilaterals*. Illustrated by Ted Evans. New York: Marshall Cavendish, 1992.

ISBN 1-85435-459-0 $17.95

★★ Single concept, 4–8

Exploring open and closed figures begins this investigation of quadrilaterals. To check their comprehension, readers are asked to identify quadrilaterals in a large set of shapes. An excellent opportunity to investigate different shapes is the task of finding all possible quadrilaterals on a three-by-three dot grid. This investigation is followed by definitions of more specific quadrilaterals—parallelograms, rectangles, squares, rhombuses, trapezoids, kites, and arrowheads. (Since arrowheads and kites have the same definition, although they look different, it is unclear why the author introduces both terms.) Different properties of quadrilaterals are explored through sorting tasks, which include using a flow chart and building various tables to record results from investigations that explore angles, diagonals, and sides. Although common properties among the shapes can be found in the tables, no attempt is made to generalize relationships, such as all squares are rectangles. Procedures are given for constructing different quadrilaterals with such tools as a set square, compass, ruler, and protractor; the reader is asked to adapt these directions to draw other quadrilaterals. Investigations with angles include working with vertical, corresponding, alternate, and adjacent angles and determining the sum of the angles of quadrilaterals. Folding paper, building quadrilaterals with straws, exploring triangle rigidity, tiling, and building Fibonacci rectangles are some of the other well-defined investigations. Mathematical recreations, such as jigsaw puzzles and matchstick puzzles are interspersed. Spatial sense and problem-solving skills are needed for problems like transforming a square into a new shape by making one cut or creating a new square whose area is twice as large by moving two of the original vertices. Although the quality of the activities is inconsistent, *Quadrilaterals* contains some very good problem-solving activities.

————. *Ratio and Proportion.* Illustrated by Ann Baum. New York: Marshall Cavendish, 1995.

ISBN 1-85435-776-X $17.95

☆ Single concept, 4-8

Concepts of ratio are introduced in the context of producing four toys composed of eight identically shaped parts that differ only in their combinations of color. For example, two toys are composed of red and yellow parts, but the ratios of red to yellow are 1 to 1 and 3 to 1. Good problems are posed about the differences among the ratios and the toys that could be made from different collections of parts. For a number of problems, the reader is to match pictures with ratios; both the order of the ratio and size are included. Finding equivalent ratios is involved in halving the recipe for Horrible Hag's Hotpot as well as in deciding how to mix quantities of two colors to produce a new color. Other interesting applications of equivalent ratios involve assembling tables that need to be packed in cartons of 1 table top : 4 legs : 12 screws and analyzing eight-by-ten grids formed with different patterns of blue and white squares. The latter activity is extended to other grids and shapes. Similar figures are informally introduced by analyzing irregular shapes. Problems are posed that involve linear measures, scale drawings, and the concept that similar figures have proportional sides. Applications of proportions also include nonexamples; for example, if a person weighs 126 pounds at 16 years of age, what would that person weigh at 48 years of age?

————. *Shapes.* Illustrated by Ted Evans. New York: Marshall Cavendish, 1993.

ISBN 1-85435-464-7 $17.95

★★ Single concept, 4–8 (3–8)

In the opening pages, shapes are defined as being only two-dimensional. Three-dimensional figures are referred to as solids, but drawings of three-dimensional figures are called shapes. These concepts are extended to investigate perspective of everyday objects as well as of geometric shapes; a sphere's shape will always be a circle, whereas a cylinder could look like a circle or a rectangle. Three perspective investigations of ellipses involve rotating a circle, viewing cylinders, and drawing patterns by using concentric circles. A number of concepts, such as closed shapes, regular polygons, interior angles, and diagonals, are introduced and defined in activities that range from practicing skills to solving problems. "These are ... these are not" definitions are effectively used to introduce polygons. In a beginning activity a question is posed about convex shapes, but the term is not defined until page 22; this inconsistency is not a great problem, since the definition can be found in the glossary. For exploring points and segments in a problem-solving mode, the game of sprouts is presented. For practicing identifying names of polygons, a crossword puzzle is included. For exploring polygons with the greatest number of sides on different-sized arrays, a problem-solving activity is proposed. Constructions of regular polygons are investigated by using central angles and, later, by using interior angles. A guided explanation for determining the total number of degrees in a polygon is given through diagrams and text; connections are made between the ancient Babylonians and the choice of 360 degrees as the number of degrees in a circle. Tangrams are used to solve shape puzzles. The diagrams for constructing tangrams are good, but in the written instructions for steps 4 and 5, it is unclear if the reader is to estimate or devise a strategy for drawing a vertical segment and a parallel segment. Final investigations include folding paper to form a flexigon and pointed stars, tying knots in a strip to form a pentagon, and exploring Möbius strips. Overall, this interesting set of shape-related investigations involves worthwhile tasks. It should be noted that some unnecessary references are made to Christian holidays.

————. *Shape Patterns*. Illustrated by Ted Evans. New York: Marshall Cavendish, 1993.

ISBN 1-85435-465-5 $17.95

☆ Single concept, 4–8

Lots of square and triangular dot and graph paper is needed for the many investigations of patterns with shape. Drawing and coding patterns on graph paper until they repeat is a good opening activity that can lead to computer investigations with Logo. A second investigation involves distorting figures by drawing the same figure on different-sized and different-shaped grids. Different shapes and colors are used to create linear and two-dimensional patterns; both graph and triangular dot paper are needed because readers are asked to extend the patterns as well as to devise their own. Pentominoes are defined by a "this is … this is not" definition; after the reader finds the remaining shapes, several games and tasks are proposed that involve the pentominoes. Translations, rotations, and reflections are defined and introduced by moving a blue and white square to create unique designs. In a description of a rotation, the design is said to have rotated 90 then 180 degrees, but the points of rotation are not given. Tasks involving rotational symmetry also introduce order. Investigations for reflections define and explore lines of symmetry and symmetrical shapes with respect to everyday objects and such geometric figures as parallelograms, equilateral triangles, and circles. Paper cutting, tracing, and painting are also used to explore reflections. The final investigations involve tessellations or tiling; other tasks are designed to determine which shapes will tessellate and why. Color is effectively used to show how the same tessellation can change in appearance. Readers should enjoy investigating tessellations as well as the other activities described in *Shape Patterns*.

———. *Solids*. Illustrated by Ted Evans. New York: Marshall Cavendish, 1993.

ISBN 1-85435-469-8 $17.95

★★ Single concept, 4–8

Essential for exploring the many investigations throughout the book is compiling the needed materials listed in the opening pages. One beginning activity is to determine how to fold a box from a square piece of paper; three different solutions are included in the answer section. Readers are asked to find and draw the different pentominoes and then to decide which ones can be folded into an open box. Later spatial challenges involve determining which face of the pentomino will be on the bottom and finding which hexominoes can be folded into a cube. These tasks provide a good introduction to cubes, cuboids, and nets. Cutting cubes to form half cubes introduces prisms. Nets for dissections of regular tetrahedrons and cubes are also given; they form interesting puzzles and can help readers develop insight into relationships among shapes. Other investigations include designing a net for an oblique cuboid and tetrahedrons and folding nets to construct such shapes as rhombic dodecahedrons, pyramids, pentagonal dodecahedrons, and stellated dodecahedrons. The shapes are intriguing and challenging!

Puzzle pieces made from smaller cubes are used to construct a larger cube; also, all combinations of stacking four cubes together are explored. The classic problem of painting a $3 \times 3 \times 3$ cube to determine the number of smaller cubes that have zero, one, two, or three faces painted is posed and extended to a $100 \times 100 \times 100$ cube. Additionally, directions are given for braiding solids and analyzing Schlegel diagrams. Throughout the book, cubes, prisms, and polyhedrons are referred to as solids rather than three-dimensional shapes, which incorrectly implies that the shape comprises the faces *and* its interior. Pyramids and cones are not well defined. Faces and edges appear to include curved edges, since "Cylinders are prisms that have circular ends" (p. 18), and a sphere is considered to have one face and one edge. "Octahedrons Mobiles" may be more appropriately titled "Decorating with Octahedrons." In spite of these inconsistencies, *Solids* can be recommended for its excellent collection of investigations.

———. *Statistics*. Illustrated by Ted Evans. New York: Marshall Cavendish, 1993.

ISBN 1-85435-468-X $17.95

☆ Single concept, 4–8

From using simple tally charts to constructing pie graphs and to calculating averages, the focus of *Statistics* is on the procedures involved in displaying data. Two activities involving tally charts are described; the detailed instructions include pictures of a sample that shows how to tally in groups of five and how to construct a chart.

Readers explore procedures for constructing pictographs and explanations of their limitations with the same data they used for the tally exercises. An interesting introduction to bar graphs investigates the composition of the front page of a newspaper with respect to headlines, pictures, and text. Titles, labels, and scales for bar graphs are discussed, and examples of mixed scales are included for the reader to analyze. Detailed directions for pie charts include how to measure angles and how to calculate degrees and percentages. Procedures for finding medians, means, and modes are presented with some questions on which best represents a given set of data. The discussion of line graphs includes guidelines for deciding when line graphs are appropriate. Some of the problems with constructing a survey are discussed, but as noted throughout the review, *Statistics* is designed to introduce procedures rather than to collect and analyze data.

————. *Time, Distance, and Speed.* Illustrated by Ted Evans. New York: Marshall Cavendish, 1993.

ISBN 1-85435-467-1 $17.95

☆ Single concept, 4–8

"A band of thirty musicians takes five minutes to play a dance. How long will a band of forty-five musicians take?" This humorous puzzle introduces the section on time. Connections to history are made in the explanations of time measures and instruments like obelisks, sundials, and calibrated candles. Of special interest is the section explaining why the year at the end of the century is usually not a leap year but why the year 2000 will be. Some of the activities are in puzzle format but simply require reading clock faces; other activities use problem solving and a mystery. One investigation deals with how long it would take to count to a million. Distances between towns on a map are calculated with scales, and then a map mileage chart is filled in. Examples involving constant speeds, average speeds, and units of speed are presented. Units like knots, the Beaufort scale, and mach numbers are explained. Coordinate graphs are used to record speed and are extended for interpreting the travel story from the graph.

————. *Triangles.* Illustrated by Ted Evans. New York: Marshall Cavendish, 1993.

ISBN 1-85435-461-2 $17.95

☆ Single concept, 4–8

Triangles is a blend of skill exercises and problem-solving activities. Many concepts found in the middle-grades curriculum are presented, but because of the many definitions given and the questions that require recognition skills, parts of the book resemble a traditional textbook. Activities and puzzles added to these sections lighten the tone; other sections are definitely investigations.

Beginning activities include folding right angles, drawing and analyzing triangles drawn on a curved surface, and exploring triangles that can be formed when one side is the circle's diameter and the third vertex is on the circle's circumference. To determine how many triangles are embedded in a figure or to solve a matchstick puzzle, spatial sense is essential. Some of the construction patterns involving circles and triangles are intriguing, and many conjectures are to be explored. Equilateral and isosceles triangles are introduced by showing several examples; the reader is to analyze the differences in sides and angles to determine appropriate definitions and characteristics for each classification. Definitions continue with acute, obtuse, right, and reflex angles; one activity has readers identify and color different types of triangles. The number of degrees in a triangle is informally explored by tearing and rearranging the angles to form a half circle. This is followed by a proof that uses parallel lines and corresponding, vertical, and alternate angles and an exercise to find the third angle from limited given information about the triangle. Traditional constructions (side, side, side; angle, side, angle; and side, angle, side) are presented as algorithms. Some questions are designed to result in impossible constructions or multiple answers. Two motivating tasks include using dissection puzzles for illustrating the Pythagorean theorem and analyzing and drawing the Penrose triangle, an optical illusion.

Math Counts

Full-page color photographs of children involved in exploring mathematics dominate this beautifully designed series. Their involvement in simple activities and everyday applications helps develop each topic. The conversational text, superimposed on each illustration, invites the reader to participate in the activities as well as to become more aware of the mathematics in the world around us. Originally the series was titled Knowabout and was published by Franklin Watts in 1988, first in Great Britain and later in the United States. The colorful covers are designed to appeal to the young child. An introduction in each thirty-two-page book discusses the importance of children seeing, talking, touching, and experimenting to learn mathematical ideas. It is noted that the photographs and text have been designed to encourage discussion between adult and child. Both children and adults of various races are shown throughout the series. An index is included.

Pluckrose, Henry. *Capacity*. Illustrated by Chris Fairclough. Chicago: Children's Press, 1995.

ISBN 0-516-05451-1   (library binding) $17.80
ISBN 0-516-45451-X   (paper)   4.95

★★★ Single concept, 1–5

Capacity is effectively introduced through brightly colored photographs of a pair of hands gently emptying firmly packed sand from a pail. The space inside a container is explored by measuring both liquid and solid materials. Questions about comparisons occur throughout the text as the reader compares familiar containers of different shapes and sizes. Although metric units are shown in the pictures, only customary units are presented in the text. Beautiful color photographs complement the simple text and invite the readers to explore capacity concepts with activities of their own.

————. *Counting*. Illustrated by Chris Fairclough. Chicago: Children's
Press, 1995.

ISBN 0-516-05452-X   (library binding) $17.80
ISBN 0-516-45452-8   (paper)   4.95

★★ Single concept, PS–6

Questions throughout the text encourage discussion about the
mathematical concept of counting. The beautiful photographs focus
expertly on familiar objects in natural settings. A pattern using
bright red and blue counters invites readers to use their own
counters to complete a matching design. This activity is followed by a
challenging review of counting activities as the reader is asked to
find how many counters, how many zoo animals, and how many
people are on one colorful spread. One-to-one correspondence, more
than, and less than are also introduced as the reader compares sets
to determine which has more.

————. *Length*. Illustrated by Chris Fairclough. Chicago: Children's Press,
1995.

ISBN 0-516-05453-8   (library binding) $17.80
ISBN 0-516-45453-6   (paper)   4.95

★★★ Single concept, 1–3

Which folded-up shoelace would be the most appropriate for each
brightly photographed shoe? This initial estimation question
promotes the definition of length—the measurement of something
from end to end. Both nonstandard (handspans and paces) and
standard (metric and customary) measurement units are introduced.
Different length units, such as centimeters and inches, meters and
feet, and kilometers and miles, appear in situations that show the
need for selecting an appropriate unit. Comparisons and the
differences between height and length are also explored. This
beautifully designed book with brilliant photographs and simple text
is typical of this series by Pluckrose and Fairclough.

————. *Numbers*. Illustrated by Chris Fairclough. Chicago: Children's
Press, 1995.

ISBN 0-516-05454-6   (library binding) $17.80
ISBN 0-516-45454-4   (paper)   4.95

★★★ Single concept, PS–2

"Can you imagine a world without numbers?" This book is
designed to encourage young children to talk about the many uses of
numbers. The full-page color photographs and the corresponding text
provide interesting and meaningful real-life contexts for this
discussion. The uses of number (ordinal, nominal, measurement)
include such examples as the numerals on a telephone, the buttons
for elevator floors, streetcars, a car license plate, house numbers, a
wristwatch, a thermometer, and a speedometer. Pluckrose and

Fairclough offer a rich variety of illustrations for the young reader, although most depict the experiences of middle-class children. Those guiding the discussion should provide opportunities for children to develop awareness of how numbers are used in their own life experiences.

————. *Pattern*. Illustrated by Chris Fairclough. Chicago: Children's Press, 1995.

ISBN 0-516-05455-4   (library binding) $17.80
ISBN 0-516-45455-2   (paper)   4.95

★★★ Single concept, 1–3

Wallpaper, tennis shoe soles, fabric, dishes, carpets, butterflies, leaves, and flowers—all these are illustrated in brilliantly colored photographs as examples of patterns in the world around us. Rotational and line symmetry are present in these patterns as well as in the examples using Legos, pattern blocks, paper cutouts, paint, or markers. The book with its simple text should help readers become more aware of the patterns in the world around us and encourage them to create original designs.

————. *Shape*. Illustrated by Chris Fairclough. Chicago: Children's Press, 1995.

ISBN 0-516-05456-2   (library binding) $17.80
ISBN 0-516-45456-0   (paper)   4.95

★★★ Single concept, 1–3

"Move your fingers around the edges of the page. What shape have you traced?" The author's first invitation to the reader to explore shape is one of many in this simple questioning text. A double-page spread displays one example each of a square, a circle, a rectangle, and a triangle. Examples of these shapes in the environment are captured in distinctive photographs of bathroom tile, decorative manhole covers, city landscapes, and construction sites. Legos, wooden triangles, and coins are models used to explore tessellations of the plane. Pictures of locks, wheels, a honeycomb, and a telephone motivate the reader to consider questions about the function of shape. The quality photographs and the simple conversational text could stimulate further discovery and discussion.

————. *Size*. Illustrated by Chris Fairclough. Chicago: Children's Press, 1995.

ISBN 0-516-05457-0   (library binding) $17.80
ISBN 0-516-45457-9   (paper)   4.95

☆ Single concept, PS–2

The concepts of big and little are represented by a variety of objects of interest to young children. Children are guided in ordering several objects by size. The limited text with the interesting full-color

312 WONDERFUL WORLD OF MATHEMATICS

photographs can encourage meaningful oral-language development in children. Size is illustrated as comparison. On successive pages, the sizes of a car, a truck, and a bus are compared to the size of an adult standing beside each vehicle. Without adult guidance the book might confuse rather than clarify. For example, four different-sized chairs are pictured on four separate pages. The same teddy bear appears on each of the pages. To judge the relative size of the chairs, the reader needs to recognize that the teddy bear is the referent for comparison.

————. *Sorting*. Illustrated by Simon Roulstone. Chicago: Children's Press, 1995.

ISBN 0-516-05458-9   (library binding) $17.80
ISBN 0-516-45458-7   (paper)   4.95
★★ Single concept, no age given

Set concepts are explored in colorful photographs in another excellent book from the Math Counts series. A pile of objects introduces the reader to sorting activities; these objects are shown on subsequent pages in different sets and subsets depending on the rule for classification. A variety of sets are presented for the reader to decide how to sort. The examples are not limited to only one rule for classification, and the reader will need to decide what the objects have in common. The book ends as it begins—with a pile of objects. The reader is encouraged to make a pile and sort in many different ways.

————. *Time*. Illustrated by Chris Fairclough. Chicago: Children's Press, 1995.

ISBN 0-516-05459-7   (library binding) $17.80
ISBN 0-516-45459-5   (paper)   4.95
☆ Single concept, PS–3

Leaning over to turn off the alarm, the little boy wipes his sleepy eyes. The text asks, "What time do you get up in the morning? How can you tell the time?" This introduction draws the reader into a photographically illustrated exploration of units of time measurement: seconds, minutes, hours, days, weeks, months, and years. The importance of timekeeping is constantly reinforced through questions and examples. Brilliantly colored photographs of children and adults of various races depict everyday references to time measurement and its use, and the conversational text challenges the reader to participate.

————. *Weight*. Illustrated by Chris Fairclough. Chicago: Children's Press, 1995.

ISBN 0-516-05460-0   (library binding) $17.80
ISBN 0-516-45460-9   (paper)   4.95
★★ Single concept, PS–2

Full-page color photographs and the superimposed text supply the stimulus for a meaningful discussion of the concept of weight. The need to know how heavy things are is presented first; then applications of the metric system are given. Several photographs show real-life applications, for example, luggage at the airport, fruit and vegetables at the market, and a scene showing that "even elephants need to be weighed sometimes." Throughout the book, the reader is asked to estimate which of two objects is heavier. The estimates are checked by using either dial or balance scales; the reader is asked to interpret the readings or results of these measurements.

MathStart

The MathStart series claims to be connected to everyday activities and related to the NCTM Curriculum Standards. The presentation and development of the mathematics concepts across the series are uneven. Authored by Stuart Murphy, the books are designed for three different levels: Level 1, for ages 3 and up, includes counting, ordering, recognizing patterns, and comparing sizes; Level 2, for ages 6 and up, includes adding and subtracting, reading timelines, estimating, and using fractions; Level 3, for ages 7 and up, is based on the previous levels and includes multiplying and dividing, building equations, and using problem-solving strategies. The minimal text in level 1 increases appropriately in the next two levels; additionally, the text in the first two levels is written in rhyming verse. The illustrations are colorful and appealing but have distinct styles reflecting different illustrators. A two-page section at the end of each book, titled "For Adults and Kids," includes suggested activities and additional children's book titles. This is especially helpful because the story line is designed to be used as a springboard into a topic.

Murphy, Stuart J. *The Best Bug Parade.* Illustrated by Holly Keller. New York: HarperCollins Publishers, 1996.

ISBN 0-06-025871-3		$14.95
ISBN 0-06-025872-1	(library binding)	14.89
ISBN 0-06-446700-7	(paper)	4.95

☆ Single concept, PS+

A flower garden is the setting for introducing size comparisons. The first friendly bug appears with the caption, "I am big." A larger one appears announcing, "I am bigger than you are," and is followed by a third one, which is the biggest. The three bugs are then illustrated in three congruent rectangles with the captions "Big.

Bigger! Biggest!!" A similar format introduces small, smaller, smallest; long, longer, longest; and short, shorter, shortest. After the introduction of each trio, the new bugs join a line of bugs for a parade. Keller's colorful creatures are pleasing and illustrate the size comparisons being made. Additional notes encourage the reader to continue to compare with toys, pets, or family members.

———. *Betcha!* Illustrated by S. D. Schindler. New York: HarperCollins Publishers, 1997.

ISBN 0-06-026768-2		$14.95
ISBN 0-06-026769-0	(library binding)	14.89
ISBN 0-06-446707-4	(paper)	4.95

★★★ Single concept, PS+ (1–3)

In *Betcha!* as in "Betcha I can tell you how much," estimation is embedded in various situations as two boys set off to enter a contest. On the city bus they estimate the number of passengers by considering the number of people in each row, the number of rows, and the number of standees. An array approach is also considered for estimating the cars in the traffic jam on the busy street. Each time one boy estimates, the second boy counts to check the estimate. To estimate the cost of the items in a window display, they use a strategy involving rounding and adjusting. The contest of estimating the number of jelly beans in a jar provides the setting for introducing a strategy of sections and layers. *Betcha!* provides a good introduction to estimation strategies because the thinking is clearly shown through text and illustrations. The simple illustrations show the two friends of different racial backgrounds as they interact with their everyday surroundings.

———. *Divide and Ride.* Illustrated by George Ulrich. New York: HarperCollins Publishers, 1997.

ISBN 0-06-026776-3		$14.95
ISBN 0-06-026777-1	(library binding)	14.89
ISBN 0-06-446710-4	(paper)	4.95

★★ Single concept, 2+

Carnival rides for eleven best friends provide the context for exploring measurement division situations and remainders. First, the friends approach the Dare-Devil ride on which two people can fit in a seat. They fill five seats, "but 1 friend is left over from our group of 11 best friends." The illustration shows the five occupied seats with one friend on the platform; an array is shown sectioned with five columns of two stars and a sixth column with one star. On the next page another child is invited along; the array is modified to show six columns of two and "12 divided by 2 = 6 full seats." Each seat on the Satellite Wheel holds three people; the situation of 11 divided by 3 is also depicted through illustrations and an array. Again, a new child joins the partial group, so 12 divided by 3 is shown. A third situation introduces 11 divided by 4 and 12 divided by 4. The final problem involves a ride for fourteen people, so the eleven best friends are joined by the other three children; this last problem is stated as $11 + 3 = 14$, not as a division problem. The group of best friends represents different ethnic backgrounds; their camaraderie, fun, and excitement are captured in Ulrich's watercolor with pen-and-ink illustrations.

————. *Elevator Magic.* Illustrated by G. Brian Karas. New York: HarperCollins Publishers, 1997.

ISBN 0-06-026774-7		$14.95
ISBN 0-06-026775-5	(library binding)	14.89
ISBN 0-06-446709-0	(paper)	4.95

★★ Single concept, 1+ (PS–1)

Ben's elevator ride provides the context for three subtraction situations. As Ben and his mother leave her tenth-floor office, she states that they need to stop at the bank two floors down. He thinks, "We're on 10. / 2 floors down / $10 - 2 = 8$," so he pushes the button marked 8. The illustration shows Ben counting down two from 10 to 8 on the panel of elevator buttons. This format is very similar to a vertical number line. The second problem is stated in parallel form, but the third problem considers the number of floors from the fourth to the first floor that Ben and his mother need to travel to meet his dad. All problems are solved similarly. The "magic" in the book's title refers to the surprise scenes that only Ben sees at each stop; the Farm Bank and Trust on the fifth floor "really" has a horse, cow, donkey, and so on. Similar scenes are found at Speedway Delivery and the Hard Rock Candy Store. Individuals of different ethnic backgrounds are present in the pleasing, simple illustrations that appear to be watercolor and pencil.

————. *A Pair of Socks.* Illustrated by Lois Ehlert. New York: HarperCollins Publishers, 1996.

ISBN 0-06-025879-9		$14.95
ISBN 0-06-025880-2	(library binding)	14.89
ISBN 0-06-446703-1	(paper)	4.95

★★ Single concept, PS+

Finding matching pairs is the focus of this Level 1 book. Colors of bright red, blue, yellow, and green and patterns of stripes and dots are depicted in different combinations of socks; the reader is to decide which socks match. Additional opportunities for matching are included on the final pages as well as suggestions for looking for other patterns and matches in the world around us. Ehlert's collages effectively combine bold color and simple lines for striking illustrations.

———. *Every Buddy Counts.* Illustrated by Fiona Dunbar. New York: HarperCollins Publishers, 1997.

ISBN 0-06-026772-0		$14.95
ISBN 0-06-026773-9	(library binding)	14.89
ISBN 0-06-446708-2	(paper)	4.95

☆ Single concept, PS+

"When I wake up feeling lonely— / crummy, yucky, very sad— / I count up all my buddies, / and I'm happy, cheery glad." Murphy's charming story in rhyming verse and Dunbar's pleasing, detailed illustrations in crayon effectively introduce the numbers from one through ten. As a little girl counts from one hamster, two sisters, three kittens to ten teddy bears, she relates how lucky she is to have all these buddies—buddies she can count on and buddies who can count on her. Each new set is shown in a scene as well as in a group beside the word name and numeral. One summary page shows all ten sets with the appropriate numeral and word names. The end pages are cleverly decorated with four rows of uniquely designed numerals from 1 through 10.

———. *The Best Vacation Ever.* Illustrated by Nadine Bernard Westcott. New York: HarperCollins Publishers, 1997.

ISBN 0-06-026766-6		$14.95
ISBN 0-06-026767-4	(library binding)	14.89
ISBN 0-06-446706-6	(paper)	4.95

★★ Single concept, 1+

Westcott's pleasing pastels illustrate a child's venture into solving a problem by collecting data. A little girl notes her family's busy life and decides that they need a vacation, but she does not know where they should go. She polls family members to find out what is important to them and records each one's responses with respect to such variables as distance, temperature, favorite activities, and whether to take the family cat. After collecting all the data, she decides that only one place fits all the conditions. The perfect vacation would be camping in their backyard. Murphy's story line is a good context for beginning data exploration.

———. *Get Up and Go!* Illustrated by Diane Greenseid. New York: HarperCollins Publishers, 1996.

ISBN 0-06-025881-0		$14.95
ISBN 0-06-025882-9	(library binding)	14.89
ISBN 0-06-446704-X	(paper)	4.95

☆ Single concept, 1+

Rich, warm colors compose the pleasing illustrations painted by Greenseid. A little girl's dog encourages her to get out of bed and get ready for the day. Time passes—five more minutes to snuggle in bed and three more minutes to wash—how much time has elapsed? Each activity is shown with a different-colored number segment that is then linked with the previous ones to find the total time lapsed. As she continues to get ready for school, different times are added two at a time: 5 and 3, 8 and 2, 6 and 7, 4 and 1. The final question asks the reader to find the total lapsed time. It should be noted that addition, not time concepts, is being developed. In the notes at the end of the book, readers are encouraged to collect their own data and construct their own time lines.

————. *Give Me Half!* Illustrated by G. Brian Karas. New York: HarperCollins Publishers, 1996.

ISBN 0-06-025873-X		$14.95
ISBN 0-06-025874-8	(library binding)	14.89
ISBN 0-06-446701-5	(paper)	4.95

☆ Single concept, 1+

Two siblings squabbling over food set the stage for *Give Me Half!* First it is pizza. Little brother offers his sister one slice, but a parent intervenes and states that they must share with each getting half. When it is time for juice, big sister offers a sip, but a parent again intervenes and says that they must share. Each time a parent intervenes, a diagram shows the pizza cut into two equal-sized parts and a can of juice filling two glasses halfway. Two sentences are written for each situation: 1/2 and 1/2 is 1 and 1/2 + 1/2 = 1. When they share two cupcakes, the text notes that 1 is 1/2 of 2. Each time the children are reminded to share, they become more frustrated; the situation erupts into a food fight. When they realize what they have done, they cooperatively share the cleanup. Although the conflict is finally resolved, the mathematics could be lost in the bickering story line. Karas's illustrations, created with acrylics, gouache, and pencil, effectively complement this introduction to fractions.

————. *Ready, Set, Hop!* Illustrated by Jon Buller. New York: HarperCollins Publishers, 1996.

ISBN 0-06-025877-2		$14.95
ISBN 0-06-025878-0	(library binding)	14.89
ISBN 0-06-446702-3	(paper)	4.95

☆ Single concept, 2+

A hopping contest between two frogs is the context for introducing addition and subtraction sentences in *Ready, Set, Hop!* Matty and Moe set out to determine who has longer hops, so they count the number of hops each of them takes to hop a specific distance. In the first contest, Moe needs five hops and Matty needs two hops more; each illustration uses loops to indicate the number of hops. A question, "How many hops did it take Matty to get to the rock?" is followed by a drawing of both Moe's and Matty's hops and then the statement "5 plus 2 equals how many? 5 + 2 = ?" Two other situations involve seven minus three and seven plus two minus one. The final problem is to determine the total number of hops for both frogs; their totals become equal as Moe takes one more hop into the pond for a "splashing" ending. The endpapers include more equations that use figures rather than symbols. Buller's cartoon characters in a woodland setting illustrate the story.

———. *Too Many Kangaroo Things to Do!* Illustrated by Kevin O'Malley. New York: HarperCollins Publishers, 1996.

ISBN 0-06-025883-7		$14.95
ISBN 0-06-025884-5	(library binding)	14.89
ISBN 0-06-446712-0	(paper)	4.95

★★ Single concept, 2+

O'Malley's lush watercolors depicting an Australian setting and Murphy's use of repetitive verse combine to tell a delightful story. When Kangaroo asks one of his friends, "Hi Emu! It's my birthday. Will you play with me?" he is answered by "Sorry Kangaroo, I have too many emu things to do." Emu is busy baking one cake, spreading two frostings, decorating with three flowers, and adding four candles. This is 1×1, 1×2, 1×3, and 1×4 for a total of 10 emu things to do! When Kangaroo makes this same inquiry to two platypuses, they each need to "slice one kiwi, squeeze two oranges, pour three cans of ginger ale, and scoop four big scoops of sherbet." This is 2×1, 2×2, 2×3, and 2×4 for a total of 20 platypus things to do! This pattern continues with inquires to three koalas and four dingoes. Each time, a summary page shows the four equations with an appropriate number of objects to represent each situation. The happy ending shows all the friends gathered for a surprise party for Kangaroo with their 10 emu things plus 20 platypus things plus 30 koala things plus 40 dingo things for 100 kangaroo things to do! Good problem situations and patterns are embedded in this delightful story with its appealing illustrations.

Measure Up with Science

Each thirty-two-page book in this six-volume set centers on the techniques and value of measurement in our everyday world. Intriguing facts and interesting stories about the history of each topic are interwoven throughout the easy-to-read text. A time line that lists important events from antiquity to current dates helps provide a perspective for various discoveries. Photographs of ancient and modern structures, nature scenes, people depicted at home or at work, and artifacts effectively illustrate the role of measurement both currently and historically. Diagrams are used to convey additional information. Throughout each book, experiments to help the reader explore various aspects of the concept are described in sections called "Something to Try." If materials are needed, a list of equipment is provided; often these are household items. A section titled "For More Information" lists more things to do, books to read, videotapes to preview, and places to visit in the United States and Canada. A table of contents and an index are helpful features.

Walpole, Brenda. *Counting.* Illustrated by Dennis Tinkler and Chris Fairclough. Milwaukee: Gareth Stevens, 1995.

ISBN 0-8368-13659-6 $17.27

☆ Single concept, (4–6)

Examined through a historical lens, the origins, uses, and types of numbers are the focus of *Counting.* From tally sticks and quipus to ancient numeration systems of the Sumerians, Egyptians, and Chinese, the history of number is documented. Photographs of a Sumerian clay tablet and Egyptian hieroglyphics and sketches of Chinese bamboo counting sticks illustrate some of the ancient symbols; diagrams convey the values and shapes of the different symbols. Readers learn that the arabic numerals we use today are based on work by Hindu mathematicians in India between 200 B.C. and A.D. 600. Fibonacci's *Liber Abaci,* published in 1202, increased their popularity, but it was not until 1500 that all of Europe was familiar with arabic numerals. References are briefly made to integers, and an explanation of the binary system is weak. The Chinese abacus, the Roman counting board, Napier's bones, and magic squares are explored in "Something to Try." Examples of numbers as ratios include golden rectangles, the Parthenon, and Fibonacci numbers in nature—the English daisy with rings of thirty-four, fifty-five, and eighty-nine petals. Plotting numbers in a coordinate plane is used as a springboard to introduce scale drawings. Technology's role is traced from Pascal's first calculator in 1642 to Burroughs's adding machines in 1892 to current computers and laptops. Through both text and a time line, readers can gain an appreciation of number as a measure and its development.

————. *Distance.* Illustrated by Dennis Tinkler and Chris Fairclough. Milwaukee: Gareth Stevens, 1995.

ISBN 0-8368-1360-X $17.27

★★ Single concept, (4–6)

Spans, fathoms, cubits, digits, and palms are some of the nonstandard units described in *Distance* as the history of linear measures is traced. A furlong was defined as the distance a horse could pull a plow without stopping to rest. As the need for standard measures increased, a royal cubit was made from black granite in about 3000 B.C.; a foot length was standardized as the length of a Roman centurion's sandal. The evolution of the Imperial system and the metric system in France in 1791 is also documented. Both measuring and estimating activities are described in "Something to Try." The relationships among such body measures as cubits, fathoms, and strides are explored. Directions are included for constructing calipers to measure circular objects and for making a measuring chain or a road measurer for long distances. Ratios are used to estimate heights of trees or to devise scale drawings and maps. Artwork and sketches effectively illustrate how distance is conveyed through perspective drawings. This thoughtful combination of photographs and other illustrations, historical facts and stories, and the activities to be tried should greatly enhance a reader's appreciation of the study of linear measures.

————. *Size.* Illustrated by Dennis Tinkler and Chris Fairclough. Milwaukee: Gareth Stevens, 1995.

ISBN 0-8368-1361-8 $17.27

★★ Multiconcept, (4–6)

Attributes explored in *Size* include area, volume and capacity, and mass and weight. A rectangle divided into square centimeters introduces the area formula for rectangles. Diagrams of small sectors of a circle that are rearranged to form a rectangle illustrate the circle's area formula as radius times half the circumference. In

"Things to Try," problems involving irregular areas like the surface of a leaf or a hand print are posed. Larger areas, such as acres and hectares, are described and pictured. Volume activities include finding volumes of different-sized cubes. The story of Archimedes is connected to experiments using water displacement to find the volumes of irregular shapes. Liters, as well as cups, pints, and quarts, are described in a section on capacity. Distinctions between mass and weight are clearly explained. In addition to estimating weights, activities include directions for devising and using balance scales, spring balances, and top pan balances. Connections to our world include the variations in size caused by weather changes that need to be accommodated in building bridges or designing water pipes. Curious facts and interesting stories from history are embedded throughout. One entry describes how the ancient Chinese measured grain and wine in containers that were carefully weighed and shaped and rang like a bell when struck; they used the same word for grain measure, wine bowl, and bell. Other interesting trivia are included in the time line; for example, the largest hamburger weighed 5520 pounds and was made in Seymour, Wisconsin, in 1989.

————. *Speed.* Illustrated by Dennis Tinkler and Chris Fairclough. Milwaukee: Gareth Stevens, 1995.

ISBN 0-8368-1362-6 $17.27

☆ Single concept, (4–6)

Speed, Time, and *Temperature* are more science-oriented than the other three books in this series. Rate is the mathematical emphasis of this book, and the reader has a number of opportunities to collect and analyze data on speed. Some of these include measuring walking and running rates or recording the changes in heart rates before and after exercise. Science is investigated by studying speed in different environments, such as speed on land versus on water or in the air. Directions are given for building devices to measure the direction and force of the wind. Some experiments involve friction and ball bearings; others involve the effects of streamlining objects on speed. Explanations and experiments on the speeds of light and sound are given; the relationship between these measures is used to estimate how far away lightning is when thunder is heard.

————. *Temperature.* Illustrated by Dennis Tinkler and Chris Fairclough. Milwaukee: Gareth Stevens, 1995.

ISBN 0-8368-1363-4 $17.27

☆ Single concept, (4–6)

Like *Speed* and *Time, Temperature* is more science-oriented than the other three books in this series. Experiments include examining perceptions of temperature as well as measuring temperatures. A history of thermometers is given as well as directions for making one. Temperature changes during the day and seasonal changes are extended to explanations of our solar system. Data collection is involved in "Something to Try" as readers are encouraged to keep a temperature diary, to measure how fast heat travels through different objects, and to study the effects of various types of insulation. Data are also collected in measuring temperature changes on plant growth, but the main purpose is the scientific study of the *effect* of temperature. Similarly in the comparisons of the energy consumptions and blood temperatures of different animals, the main purpose is the study of science, but number is essential to the science investigation.

————. *Time*. Illustrated by Dennis Tinkler and Chris Fairclough. Milwaukee: Gareth Stevens, 1995.

ISBN 0-8368-1364-2 $17.27

★★ Single concept, (4–6)

Understanding the various ways to measure time is one of the main themes of *Time*. "Something to Try" includes directions for building shadow clocks, water clocks, and sand clocks as well as for using candles, gears, and pendulums to measure time. Time measures are part of the scientific explanation of day and night and the changing seasons. These time periods evolve into a discussion on the history of calendars, both the Gregorian calendar and the Chinese and Islamic lunar calendars. Conflicts and adjustments that needed to be made to the calendar because of inconsistent measures are intriguing. Some of this conflict is documented in the illustration of Hogarth's painting *Give Us Back Our Eleven Days*. Twenty-four-hour clocks are explained as well as their advantages for recording train and airplane timetables and dealing with schedules across different time zones. Readers are also encouraged to construct their own autobiographical time lines and to collect and chart data on their activities in the course of a day.

Author Index

Title Index

(Titles having numerals are alphabetized as if spelled out.)

341